T0269065

Hydrogen Electrochemical Production

Hydrogen and Fuel Cell Primers

Hydrogen Electrochemical Production

Christophe Coutanceau
Institute of Material and Environment of Poitiers (IC2MP),
University of Poitiers, Poitiers, France

Stève Baranton
Institute of Material and Environment of Poitiers (IC2MP),
University of Poitiers, Poitiers, France

Thomas Audichon
Institute of Material and Environment of Poitiers (IC2MP),
University of Poitiers, Poitiers, France

Series Editor
Bruno G. Pollet
Department of Energy and Process Engineering
Norwegian University of Science and Technology,
Trondheim, Norway

ACADEMIC PRESS

An imprint of Elsevier

Academic Press is an imprint of Elsevier
125 London Wall, London EC2Y 5AS, United Kingdom
525 B Street, Suite 1800, San Diego, CA 92101-4495, United States
50 Hampshire Street, 5th Floor, Cambridge, MA 02139, United States
The Boulevard, Langford Lane, Kidlington, Oxford OX5 1GB, United Kingdom

Notices
Knowledge and best practice in this field are constantly changing. As new research and experience broaden our
understanding, changes in research methods, professional practices, or medical treatment may become
necessary.

Practitioners and researchers must always rely on their own experience and knowledge in evaluating and using
any information, methods, compounds, or experiments described herein. In using such information or methods
they should be mindful of their own safety and the safety of others, including parties for whom they have a
professional responsibility.

To the fullest extent of the law, neither the Publisher nor the authors, contributors, or editors, assume any
liability for any injury and/or damage to persons or property as a matter of products liability, negligence or
otherwise, or from any use or operation of any methods, products, instructions, or ideas contained in the
material herein.

British Library Cataloguing-in-Publication Data
A catalogue record for this book is available from the British Library

Library of Congress Cataloging-in-Publication Data
A catalog record for this book is available from the Library of Congress

ISBN: 978-0-12-811250-2

For Information on all Academic Press publications
visit our website at https://www.elsevier.com/books-and-journals

  Working together
to grow libraries in
developing countries

www.elsevier.com • www.bookaid.org

Publisher: Joe Hayton
Acquisitions Editor: Raquel Zanol
Editorial Project Manager: Mariana Kuhl
Production Project Manager: Paul Prasad Chandramohan
Cover Designer: Victoria Pearson

Typeset by MPS Limited, Chennai, India

CONTENTS

Bruno G. Pollet is a full Professor of Renewable Energy at the Department of Energy and Process Engineering at the Norwegian University of Science and Technology (NTNU) in Trondheim. His research covers a wide range of areas in electrochemical engineering, electrochemical energy and Sono-electrochemistry (the use of Power Ultrasound in Electrochemistry) from the development of novel materials, hydrogen fuel cell to water treatment/disinfection demonstrators & prototypes. He was a full Professor of Energy Materials and Systems at the University of the Western Cape (South Africa) and R&D Director of the National Hydrogen South Africa (HySA) Systems Competence Centre. He has worked for Johnson Matthey Fuel Cells Ltd (UK) and collaborated with many key hydrogen and fuel cells industries worldwide. He was awarded a Diploma in Chemistry and Material Sciences from the Université Joseph Fourier (France), a BSc (Hons) in Applied Chemistry from Coventry University (UK) and an MSc in Analytical Chemistry from The University of Aberdeen (UK). He also gained his PhD in Physical Chemistry in the field of Electrochemistry and Ultrasound at the Coventry University Sonochemistry Centre of Excellence (UK).

Christophe Coutanceau obtained his Ph.D. degree in electrochemistry in 1994 at the University of Poitiers, France. He then worked as Assistant Professor at the Laboratory of Catalysis in Organic Chemistry (LACCO) in that same institution, before being promoted in 2008. At present, he is Professor in Physical Chemistry at IC2MP and Director of the Catalysis and Unconventional Media group of the Institute of Chemistry of Poitiers. He is also chair of the Low Temperature Fuel and Electrolysis Cells axis of the Hydrogen-Systems-Fuel Cells Research Grouping from the French National Council of Scientific Research (CNRS). For more than 20 years, his research interests include development of synthesis methods for nanostructured supported metals using green processes (microwave activation, electrochemical methods, etc.) and solvents (polyols, water) and their application as electrocatalysts in fuel cells, electrolysis cells, and electrosynthesis reactors. He is also interested in the valorization of agroresources using electrochemical methods. He has presented or published over a hundred articles in reviewed international journals, as well as several book chapters, invited lectures, and world patents related to fuel cells, electrolysis cells, and biomass conversion.

Stève Baranton obtained his Ph.D. degree in electrochemistry in 2004. He worked conducted research in Canada and Japan before joining IC2MP as Assistant Professor. His current research interests include the development of synthesis methods for nanostructured supported metals using green processes (microwave activation, electrochemical methods, etc.) and solvents (polyols, water) and their application as electrocatalysts in fuel cells, electrolysis cells, and electrosynthesis reactors. He also develops in situ spectroscopic methods. He has presented or published over 90 documents related to fuel cells, electrolysis cells, and biomass conversion, including articles in reviewed international journals, book chapters, invited lectures, and world patents.

Thomas Audichon obtained his Ph.D. degree in Materials and Electrochemistry in 2014 from the University of Poitiers, France. He has a postdoctoral position at the Institute of Material and

Environment of Poitiers (IC2MP). His current research interests include synthesis and characterization of nanomaterials (oxides and metals) for oxygen reduction reaction and anodic activation of water in electrolysis cells, and formulation of membrane-electrodes assemblies and measurement of electrochemical performances on an electrolysis test bench. He published several papers in international peer reviewed journals on materials for water electrolysis cells.

CHAPTER 1

Hydrogen Electrochemical Production

1.1 INTRODUCTION

Since the beginning of the industrial revolution, in the middle of the 19th century, fossil fuel resources have been intensively exploited in order to cope with the increasing demand of energy and raw materials for mass production. Moreover, the exponential growth of the human population and the economic development of emergent countries induce a strong pressure on the demand of energy and goods, which leads to an increase in the consumption of fossil fuels. But the amount of these vital resources on earth, such as coal, natural gas, and hydro-carbons, is limited, and the natural cycle of their formation runs over a geological period of time. As a consequence, their availability will inexorably decrease. Some projections forecast that fossil fuels could be exhausted in few decades. Moreover, the intensification of human industrial activity and the development of transportation means have a very high impact on the greenhouse gas emission in the atmosphere. Therefore, the use of a more efficient and less polluting energy vector has been proposed to replace fossil fuels, leading to the emergence of the Hydrogen Economy concept. The success of this approach relies heavily on the broad availability of hydrogen that can be produced either by thermal reforming of hydrocarbons or biomass, electrolysis of water or electroreforming of biomass. In this context, the two main areas of hydrogen use are in the chemical industry (e.g. catalytic processes for fuels, chemicals requiring hydrogen etc.) and the production of electricity through fuel cell systems [1,2].

Currently, ca. 60 million tons (ca. $650\,G\,N\,m^3$) of hydrogen are produced per year, and its consumption increases by ca. 6% per year. [3]. The main industrial uses of hydrogen are for the refinery of fossil oils (ca. 50%) and the synthesis of ammonia (ca. 34%), the remaining representing ca. 16%, is used for the synthesis of other chemicals or for energy production [4]. In oil refinery, hydrogen is first used to increase the H/C ratio in order to improve the quality of fuels and

Hydrogen Electrochemical Production. DOI: http://dx.doi.org/10.1016/B978-0-12-811250-2.00001-7

second to decrease the impurities (oxygen, nitrogen, and sulfur) in crude oils through hydrotreatment reactions. For example, the sulfur content in fuels has to be decreased from several hundreds of ppm in crude oils to less than 10 ppm in refined oils in order to fulfill environmental standards. Hydrogen is extensively used for the production of ammonia which in turns is used as fertilizer. The intensification of the agriculture to feed the growing human demography explains the high production volume of this compound and the increase consumption of hydrogen. The third class of industrial applications of hydrogen consists in the hydrogenation reactions of compounds to obtain fine chemicals in petrochemical industries. The use of hydrogen for energy remains still anecdotic. But recently, several car manufacturers started to produce and to commercialize fuel cell cars (Toyota Mirai and Hyundai ix35 Fuel Cell, Nissan, General Motors, Daimler, etc.). The need for hydrogen will then increase to supply the growing fuel cell vehicle fleet, and hence the intensification of its production.

Hydrogen is one of the most abundant elements in the universe, but in the earth's crust, it only represents 0.15 at%. Native dihydrogen cannot be found easily in significant amount, and the available hydrogen is almost always combined with other elements, such as carbon in hydrocarbons, oxygen in water, and both in biomass. As a consequence, hydrogen has to be produced from those raw resources. Currently, 96% of hydrogen is produced from fossil fuels and 4% from electrolysis [5], because reforming of hydrocarbons is a well-mastered industrial process and still remains less costly than water electrolysis. Amongst the fossil fuels for hydrogen production, natural gas represents ca. 50%, followed by oil (ca. 30%) and coal (ca. 20%) [6]. On the other hand, fossil fuel reforming produces greenhouse gases and hydrogen containing significant amounts of carbon monoxide (CO). [7] Thermal treatment of biomass can also lead to the formation of H_2/CO reformate, without the problem of greenhouse gas emission. In this case, the H_2/CO syngas can be directly used for the production of green hydrocarbons through the Fisher–Tropsch process. But most of chemical processes in the industry need high-purity hydrogen, as for example the synthesis of ammonia. And it is also well known that the catalyst used as anode materials in low-temperature fuel cells is very sensitive to the presence of carbon monoxide (CO). Indeed, the proton exchange membrane fuel cell (PEMFC), which is the most mature fuel cell technology, uses anodes containing platinum because of its ability to oxidize H_2 to H^+. CO in H_2/CO reformate strongly adsorbs on Pt surface poisoning it [8]; as a consequence, the fuel cell performance is dramatically lowered [9].

However, hydrogen has a very high theoretical energy density, i.e., 32.9 kW h kg^{-1}, which makes it interesting for energy storage and conversion applications. But it has to be stored, either as a pressurized gas (200 to 700 bars) or in the liquid state (at 20K) greatly reducing the volumetric energy density in the range from ca. 1.5 to ca. 12 wt.% storage density (without considering the energy needed for its compression or liquefaction). The second interest of hydrogen/air fuel cells is that under reversible conditions such systems do not follow Carnot's theorem. The general reaction occurring in fuel cells is the use of hydrogen and oxygen in the presence of a catalyst according to the following equation:

$$H_2 + \tfrac{1}{2} O_2 \rightarrow H_2O \tag{1}$$

The theoretical electric energy (W_e) produced by a fuel cell under standard conditions is expressed as follows:

$$W_e = nFU^0_{cell} = -\Delta G_{f,H_2O}^0 \tag{2}$$

where U^0_{cell} is the equilibrium cell voltage under standard conditions, n is the number of exchanged electrons, F is the Faraday constant (96,485 C mol$_{H_2}^{-1}$), and $\Delta G_{f,H_2O}^0$ is the Gibbs energy of water formation (-237.1 kJ mol^{-1}).

The theoretical energy efficiency, ε_{rev}, of the fuel cell is defined as the ratio of the produced electrical energy ($-\Delta G_{f,H_2O}^0$) to the chemical combustion energy ($-\Delta H_{f,H_2O}^0 = -285.8$ kJ mol$_{H_2}^{-1}$):

$$\varepsilon_{rev} = \frac{\Delta G_{f,H_2O}^0}{\Delta H_{f,H_2O}^o} = 83\% \tag{3}$$

This value of the reversible efficiency is twice as high as that generally calculated from the Carnot's theorem for internal combustion engine, i.e., ca. 40%.

Water electrolysis can become a promising mean to produce clean hydrogen, and to further increase its availability as energy vector or industrial compound. In comparison with thermochemical water splitting and photoelectrochemical water splitting, water electrolysis is a more mature technology owing to the wide research activity devoted to the development of materials in order to improve the system efficiency, so that this technology starts to be commercialized. Nevertheless, principally owing to the high cost of the technology, hydrogen production via water electrolysis represents only 4% of the actual global hydrogen production [11]. However, this technology is often described as being the most promising one to valorize and store the electrical surplus produced by renewable energy systems (e.g. solar PV, hydraulic power stations, and wind turbines, etc.) or by electric power plants. The coupling of decarbonized energy with the production of hydrogen by electrolysis could allow developing alternative economic sectors for energy storage and conversion. Indeed, hydrogen produced by water electrolysis coupled with renewable energy systems is expected to be the best mean to develop an eco-friendly integrated system which could play a key role in the increase of green energy production in the energy mix. The advantage of this technology is the high efficiency for the production of hydrogen with high purity and without concomitant production of CO_2 and CO. The electricity restitution by hydrogen conversion can occur in a PEMFC (Fig. 1).

As explained below, the PEMFC technology requires pure hydrogen in order to avoid a rapid and drastic decrease in performance due to the occurrence of the strong absorption of by-products (CO, sulfur, etc.), which contaminates the catalysts in the electrodes. Because the only by-product formed in a water electrolysis cell is pure oxygen and both the anodic and cathodic compartments are separated by a membrane, this technology is very convenient for the production of highly pure hydrogen for energy conversion in fuel cell or chemical industry applications. Moreover, oxygen could also be stored and valorized. In

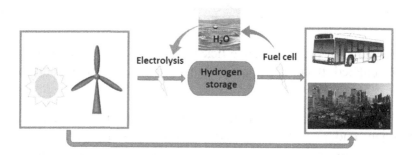

Figure 1 Hydrogen cycle (production and using) based on water electrolysis, as energetic vector [10].

the case of stationary systems, the storage of pure oxygen produced by water electrolysis via renewable energy sources allows its reuse in a fuel cell for the production of electricity with higher efficiency. Pure oxygen can also be valorized for medical care and chemical industry, in turns reducing the actual cost of water electrolysis systems.

However, the working voltage of a water electrolysis cell is close to 2.0 V, which leads to an energy consumption for hydrogen production higher than 50 kW h kg^{-1}; the theoretical value under standard conditions being 33 kW h kg^{-1}. According to thermodynamic data (Table 1), the oxidation of many alcohols leads to very low reversible potentials, close to that of hydrogen. It can then be expected that their oxidation at the anode in an electrolysis cell could dramatically decrease the cell voltage, and further save energy for the production of pure clean hydrogen (Fig. 2). An increase of the alcohol electrolysis voltage up to 1 V would still represent 50% of energy saving compared with water electrolysis and a 60% electrical yield between the production of hydrogen at 1 V and its consumption as a fuel in a PEMFC working at 0.60 V (against ca. 30% to 40% in the case of hydrogen production in a water electrolysis). This yield does not take into account the consumption of alcohol, which obviously could impact on the global process efficiency. But the use of a by-product or waste from another industrial process, such as glycerol in the case of the bio-fuel industry, and the simultaneous production of value-added chemicals annihilate the problem of the alcohol production. Indeed, the methanolysis reaction of vegetable oils leads to ca. 10 kg of glycerol

Table 1 Thermodynamic Data [15, 16] (Given in kJ per Mole of Reactant), Related Electrolysis Cell Standard Voltage (U^0_{cell}), Thermoneutral Cell Voltage (U^{th}_{cell}), and Number of Moles of Hydrogen Produced by the Electrolysis Reactions per Mole of Reactant Considering the Complete Oxidation of Water and Alcohols (Under Liquid Phases) Into Oxygen and Carbon Dioxide, Respectively

	ΔG^0_r (kJ mol^{-1})	ΔH^0_r (kJ mol^{-1})	U^0_{cell} (V)	U^{th}_{cell} (V)	n_{H_2} (mol)
Water	237.1	285.8	1.229	1.481	1
Methanol	8.9	131.0	0.015	0.226	3
Ethanol	97.5	348.1	0.084	0.301	6
Ethylene glycol	5.0	236.2	0.006	0.245	5
Glycerol	4.0	342.9	0.0029	0.257	7
Glucose	−27.4	627.1	−0.012	0.271	12

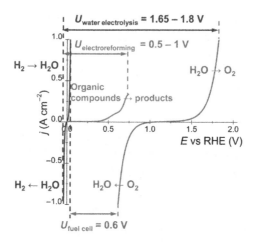

Figure 2 *Comparison of the theoretical E(j) electric characteristics representative of the Butter–Volmer kinetics law for water oxidation, alcohol oxidation, oxygen reduction, and proton reduction. ($U_{fuel \; cell}$) is the cell voltage for hydrogen/oxygen fuel cell at a current density of 1 A cm^{-2}; ($U_{water \; electrolyzer}$) and ($U_{alcohol \; electrolyzer}$) are the cell voltages for water electrolysis and alcohol electrolysis at a current density of (A) 1 A cm^{-2} [12].*

for the production of ca. 100 kg of biodiesel [13], i.e., 10 wt.%, whereas the production of bioethanol through a fermentation process leads to ca. 4 wt.% of glycerol as by-product [14], making crude glycerol a waste from the biofuel industy.

Because it is the simplest alcohol and it displays high reactivity, methanol was first studied for such an application [17]. However, it is a toxic compound which is mainly produced from fossil fuels (partial oxidation of methane). The oxidation of ethanol, which can be produced from biomass, has also been studied in acidic electrolysis cell, but it is less reactive and the C–C bond is difficult to break, which limits the cell performance and efficiency of the system [18]. Although alcohols are more reactive in alkaline medium than in acidic medium. The alkaline medium has also the advantage to tolerate anode materials for alcohol oxidation reaction with low platinum content [19] or without platinum [20]. Moreover, the ability of catalysts for the C–C bond breaking are lowered in such medium. This property can be beneficial for the coproduction of both hydrogen and value-added products from polyols produced from biomass, such as glycerol.

CHAPTER 2

Hydrogen Production From Thermal Reforming

2.1 PRODUCTION FROM HYDROCARBONS

The main industrial process for the hydrogen production consists in methane reforming, coming from either natural gas or biogas. The following two reactions are considered for the production of hydrogen from methane:

$$CH_4 + H_2O \rightarrow CO + 3\,H_2 \qquad (4)$$

$$CH_4 + 2\,H_2O \rightarrow CO_2 + 4\,H_2 \qquad (5)$$

Both reactions are very endothermic, with standard enthalpy at 25°C of $+206\,kJ\,mol^{-1}$ and $+165\,kJ\,mol^{-1}$, respectively, so that the methane reforming has to be performed at very high temperatures (close to 1000°C) over heterogeneous catalysts.

The first step of methane reforming consists in an autothermal reforming at ca. 1000°C over a ceramic as catalyst in the presence of oxygen, followed by a steam forming at ca. 900°C over a NiMgAl oxide catalyst in the presence of water steam. These reactions lead to a gas mixture containing about 20% of carbon monoxide (CO).

This gas mixture then needs to be purified, and particularly the amount of CO has to be drastically decreased to several ppm in order to be further used either in fuel cells or in the chemical industry. This step of purification is very complicated and starts with the water gas shift (WGS) reaction:

$$CO + H_2O \rightarrow CO_2 + H_2 \qquad (6)$$

Because this reaction is exothermic, with a standard enthalpy at 25°C of $-41\,kJ\,mol^{-1}$, it has to be performed at lower temperatures than reforming reactions. The heat should then be exchanged between the steam-reforming reactor and the first high-temperature water gas shift (HTWGS) reactor working at ca. 350°C, and between the HTWGS and the low-temperature water gas shift (LTWGS) reactor

Hydrogen Electrochemical Production. DOI: http://dx.doi.org/10.1016/B978-0-12-811250-2.00002-9

working at ca. 200°C. After these treatments, the gas mixture usually contains H_2, N_2, CO_2, H_2O, and still ca. 5‰ (5000 ppm) CO.

The cleaning treatments can be achieved by the removal of CO_2, followed by a methanation reaction of CO. CO_2 can be extracted by acid/base reaction at room temperature with ethanolamine, as an example, according to the following reaction:

$$HO\diagdown\diagup\diagdown NH_2 + CO_2 \rightleftarrows \; O \quad NH + H_2O \tag{7}$$

Such process can lead to residual CO_2 ranging from 0.005% to 0.1% in a volume. The removal of CO and CO_2 can then be performed by methanation reactions [21]:

$$CO + 3\,H_2 \rightarrow CH_4 + H_2O \tag{8}$$

$$CO_2 + 4\,H_2 \rightarrow CH_4 + 2\,H_2O \tag{9}$$

Because these reactions are exothermic [they are the reverse of reactions (4) and (5)], they should be performed at lower temperatures and higher pressures. After these treatments, hydrogen can then be obtained with very high purity, with less than 10 ppm and 5 ppm remaining CO_2 and CO, respectively.

An alternative purification process involves the methanation of CO and CO_2 of the gas mixture obtained just after the HTWGS and LTWGS reactions, followed by pressure swing absorption cycles in order to remove CO_2 and the remaining CO. The pressure swing adsorption process consists in the adsorption of CO and CO_2 under high pressure at the surface of adsorbent materials (molecular sieves, zeolites, active carbons, etc.) and to release the adsorbed species under reduced pressure. The repetition of cycles allows the production of very pure hydrogen.

The production of hydrogen can also be performed by dry reforming of methane, according to the following equation:

$$CH_4 + CO_2 \rightarrow 2\,H_2 + 2\,CO \tag{10}$$

The standard enthalpy at 25°C for this reaction is $+247\,kJ\,mol^{-1}$; hence, it has to be performed at high temperature and low pressure. This reaction is very interesting because it allows the generation of hydrogen and the simultaneous conversion of CO_2, which can be considered now as a cheap renewable raw material. However, after this reaction is carried out, the same purification steps as in the case of steam reforming of methane have to be performed.

Lastly, biomass can also be used for the thermal production of hydrogen. Compounds from biomass, such as cellulose, hemicellulose, and lignin, have the general $C_xH_yO_z$ formula. Fig. 3 gives, as examples, one structure of each biopolymer. These biopolymers are very stable and need high temperatures to be deconstructed.

The first step consists generally in their thermal transformation into a gas mixture containing H_2, CO, CH_4, CO_2, etc. Two different processes are generally used. The first one involves flash pyrolysis at high temperatures (close to 1000°C) of the biomass, the low temperature flash pyrolysis (around 500°C) leading to the synthesis of green oils, whereas the second one is based on the gasification of the biomass at high temperatures in the presence of oxygen and water. The second step consists in the transformation of CH_4 into CO, CO_2, and H_2 through autothermal, steam reforming, or dry reforming reactions. Then, the amount of CO in the gas mixture is decreased to a few thousands ppm by HTWGS and LTWGS reactions. At this stage, the H_2/CO mixture can be used to feed a solid oxide fuel cell (SOFC). Indeed, the high working temperature of these fuel cells (ca. 800°C) and the electrode materials based on Nickel cermet make them less sensitive to the presence of CO. But to be used in a proton exchange membrane fuel cell, the amount of CO needs to be decreased to a few ppm, which can be achieved by the same steps as described for the production of pure hydrogen by methane reforming. Fig. 4 summarizes the main processes for the transformation of biomass into compounds for energy applications.

2.2 PRODUCTION FROM ETHANOL

Ethanol can be considered as a compound originating from biomass. The first generation of bioethanol is derived from starch (Fig. 5A) or directly from sugars in (Fig. 5B) cane, cereals, or beet. Starch has first to undergo acidic hydrolysis to produce glucose. Moreover, this sugar undergoes an anaerobic enzymatic fermentation to produce ethanol. The mixture is then distilled to purify the ethanol. The bioethanol

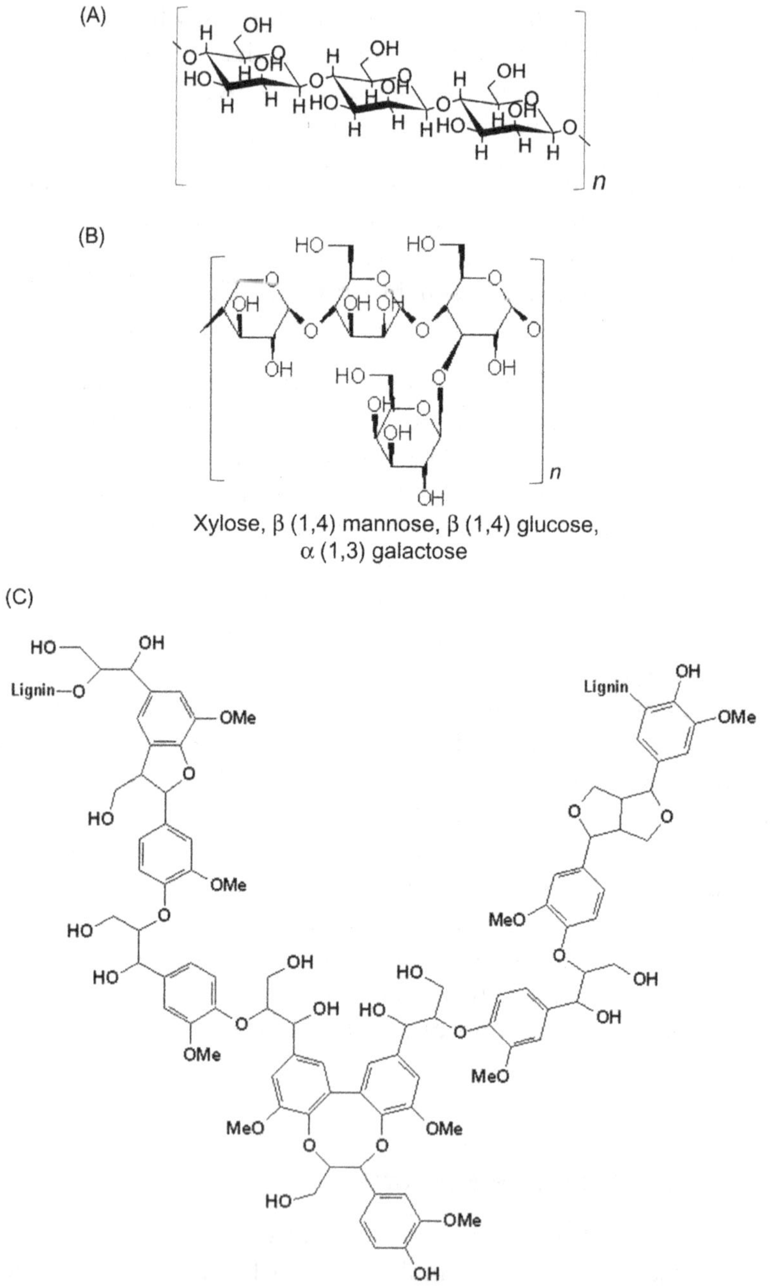

(A)

(B)

Xylose, β (1,4) mannose, β (1,4) glucose,
α (1,3) galactose

(C)

Figure 3 (A) General formula of cellulose, (B) formula of one hemicellulose [22], and (C) formula of one lignin [23].

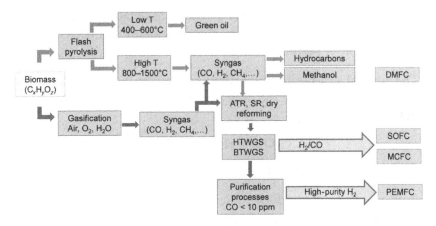

Figure 4 Schematic of the main transformation processes of biomass into compounds for energy applications.

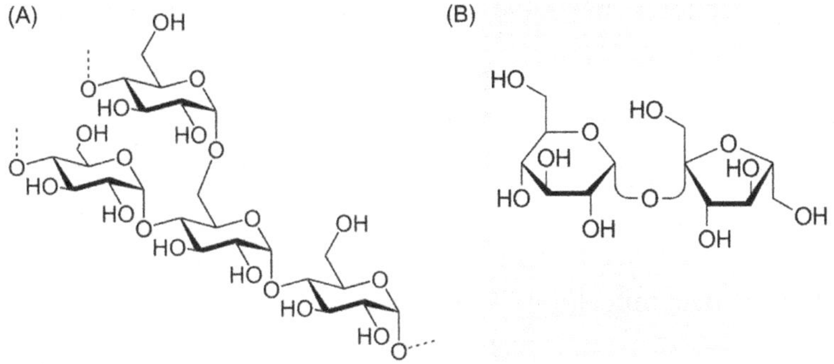

(A)

(B)

Figure 5 (A) General formula of starch [24] and (B) formula of sucrose from cane [25].

production is currently the only process being developed at industrial level. But, this process of bioethanol production is based on the transformation of edible resources and then directly competes with the alimentary needs.

The second-generation bioethanol is produced from nonedible lignocellulosic biomass [26] originating from wood and agriculture wastes (postharvest residues). But owing to the high stability of biopolymers and the complexity of the lignocellulosic biomass (composed of cellulose, hemicellulose, and lignin, see Fig. 3), their conversion into alcohol requires a pretreatment process of the biomass (e.g. steam explosion process) before the acid or enzymatic hydrolysis and fermentation/distillation processes [27], making this whole process very expensive. Third generation of bioethanol produced from enzymatic or microbial processes, or from algae is also considered, but the development of efficient processes based on this technology still needs substantial R&D before it becomes competitive.

The ethanol steam-reforming reactions leading to a higher hydrogen production (and therefore the preferred ones) can lead to the formation of CO or CO_2:

$$CH_3CH_2OH + H_2O \rightarrow 2\ CO + 4\ H_2 \tag{11}$$

$$CH_3CH_2OH + 3\ H_2O \rightarrow 2\ CO_2 + 6\ H_2 \tag{12}$$

Both reactions are endothermic, with standard enthalpies of reaction at 25°C ΔH_{298}^{0} of $+255\ kJ\ mol^{-1}$ and $+173\ kJ\ mol^{-1}$, respectively. Therefore, they have to be performed at high temperatures, i.e., under gas phase. As the number of mole increases for the steam-reforming reactions, they have to be carried out at low pressures.

However, under high-temperature and low-pressure conditions, numerous side reactions can occur, some producing hydrogen such as:

ethanol dehydrogenation: $CH_3CH_2OH \rightarrow CH_3CHO + H_2$
$$\left(\Delta H_{298}^{0} = +68\ kJ\ mol^{-1}\right) \tag{13}$$

ethanol decomposition: $CH_3CH_2OH \rightarrow CH_4 + CO + H_2$
$$\left(\Delta H_{298}^{0} = +49\ kJ\ mol^{-1}\right) \tag{14}$$

$$CH_3CH_2OH \rightarrow C + CO + 3H_2\left(\Delta H_{298}^{0} = +124\ kJ\ mol^{-1}\right) \tag{15}$$

and other reactions can occur without hydrogen production, even if they are less thermodynamically favored at high temperatures:

ethanol dehydration: $CH_3CH_2OH \rightarrow CH_2CH_2 + H_2O$
$$\left(\Delta H_{298}^{0} = +45\ kJ\ mol^{-1}\right) \tag{16}$$

ethanol hydrogenolysis: $CH_3CH_2OH + 2\ H_2 \rightarrow 2\ CH_4 + H_2O$
$$\left(\Delta H_{298}^{0} = -157\ kJ\ mol^{-1}\right) \tag{17}$$

ethanol decomposition: $CH_3CH_2OH \rightarrow 3/2\ CH_4 + \frac{1}{2}\ CO_2$
$$\left(\Delta H_{298}^{0} = -74\ kJ\ mol^{-1}\right) \tag{18}$$

$$CH_3CH_2OH \rightarrow C + CH_4 + H_2O\left(\Delta H_{298}^{0} = -82\ kJ\ mol^{-1}\right) \tag{19}$$

Duprez et al. explained that, in the case of an ethanol/water 1:1 molar ratio at 1 atm, reaction (18) occurred preferentially at low temperature and that the steam-reforming reaction started at 250°C, with a major production of CO_2 at temperatures lower than ca. 550°C and of CO at higher temperatures, the hydrogen production reached optima at 900°C [28].

In order to improve the efficiency of the ethanol steam-reforming reaction, catalysts have been developed. Two catalysts lead to relatively good performances for the ethanol-reforming reaction at 700°C, Ni/Al_2O_3 and Rh/Al_2O_3, the latter being much more active than the former one. The compositions of dry gases after reaction are very close for both catalysts, being ca. 70% for H_2, <1% for CH_4, ca. 10% for CO, and ca. 20% for CO_2. But the yield in H_2 with respect to the mass of catalytic metal is eight times higher with Rh than with Ni [29].

On Rh/Al_2O_3 catalyst, the ethanol steam-reforming mechanism is complicated because the acid and alkaline sites of the support are also involved. First, ethanol dehydrogenates on the alkaline sites of the support into acetaldehyde or into ethylene on the acid sites. Then, acetaldehyde and ethylene can be decomposed into H_2, CO, CO_2, and CH_4 on the Rh metallic active sites. Methane can further undergo a steam-reforming reaction into H_2, CO, and CO_2, whereas CO can undergo a water−gas shift reaction into H_2 and CO_2. However, ethylene can also undergo reactions leading to the formation of coke at the surface of the catalyst, leading to its rapid deactivation. In order to avoid, or at least to limit the formation of coke, an important work has been devoted to the development of new catalyst supports. Amongst the most interesting supports, CeO_2-ZrO_2 mixed oxides have shown very good stability, high activity, and lower ability for the formation of ethylene [30,31].

However, in order to avoid expensive and time-consuming purification steps, it is preferable to use raw bioethanol feed rather than pure ethanol. However, in raw bioethanol, the presence of impurities, mainly propan-1-ol, methyl-3 butanol-1, esters, aldehydes, acetic acid, and nitrogen containing bases, can have an important effect on the ethanol steam-reforming reaction. Levallant et al. [32] have shown the detrimental effect of butanol, diethylether, and ethyl acetate on the catalytic activity of $Rh/MgAl_2O_4$ catalyst: a strong deactivation of the catalyst with a decreased ethanol conversion, the decrease of the selectivity in hydrogen, and an increase in intermediate products, especially ethylene. The deactivation was then explained in terms of coke deposition onto the catalyst surface. However, the same authors have developed a $RhNi/Y_2O_3-Al_2O_3$ catalyst which led to a very high conversion of raw bioethanol into H_2, low methane ratio, and almost no coke formation on the catalyst for the steam reforming of raw bioethanol [33].

2.3 ADVANTAGES AND DISADVANTAGES

As explained earlier, 96% of the whole hydrogen production comes from fossil fuel such as natural gas, oil, and coal. The main reason is that the production cost of hydrogen by steam reforming is very low compared to that using water electrolysis. Indeed, the cost of hydrogen from fossil fuel steam reforming is less than 10 € per GJ H_2 against several tens € per GJ H_2 to several hundred € per GJ H_2 from water electrolysis when nuclear power or renewable energy sources are used, respectively [34].

But fossil fuel reserves are limited, and their use is accompanied with the production of by-products which have an impact on the environment and further on the human health. For example, the burning of fossil fuels leads to the emission of CO_2 in the atmosphere, which is a greenhouse gas, and represents one of the main contributors to the global climate change. As a consequence, the objective of the COP 21 (21st Conference Of the Parties), held in 2015 in Paris, on the world climate was to propose an universal agreement in order to decrease the greenhouse gas emission and to limit global warming between 1.5°C and 2.0°C at the end of the 21st century. In this context, the development of carbon-free energy sources, or at least the use of renewable bio-based resources has then been formally and strongly recommended.

Moreover, the thermal steam reforming of carbon-containing compounds, whatever their origin (fossil, biogas, biomass, etc.), always leads to a mixture of hydrogen, CO, carbon dioxide, methane, water, etc. Therefore, for energy applications as fuel in fuel cells or as reactant for the chemical industry, the hydrogen-containing gas mixture has to be cleaned from any impurities. For this purpose, very complex processes have to be implemented, including hydrodesulfuration and

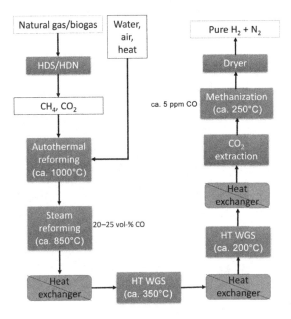

Figure 6 A schematic of the process for the synthesis of pure hydrogen convenient for feeding a proton-exchange membrane fuel cell (PEMFC) from natural gas steam reforming.

hydrodenitrogenation reactors, steam-reforming reactors, water—gas shift reactors, preferential oxidation reactors, CO_2 extraction reactors, methanation reactors, and heat exchangers, as shown in Fig. 6, for natural gas reforming as an example.

CHAPTER *3*

Hydrogen Production From Water Electrolysis

The global reaction occurring in a water electrolysis system consists in the decomposition of water molecules into dihydrogen and dioxygen molecules (Eq. 20):

$$H_2O \rightarrow O_2 + H_2 \qquad (20)$$

The water electrolysis reaction takes place in an electrochemical system that is composed of two electrodes (an anode and a cathode where oxidation and reduction of water occur, respectively) and an electrolyte (ionic conductor). The two electrodes are connected to an electric energy generator (Fig. 7). The global reaction of water electrolysis can occur under different conditions: aqueous alkaline or aqueous acidic conditions at temperatures lower than 100°C, or using solid oxide electrolytes at temperatures higher than 700°C. Depending of the electrolytes, and on the electrolysis cell working temperature, water-electrolysis cells are classified into three main categories:

- Alkaline electrolysis cell (AEC): working cell temperature $<80°C$, ionic species are hydroxyl ions (OH^-), aqueous KOH or NaOH as electrolytic media;
- Proton exchange membrane electrolysis cell (PEMEC): working temperature $<80°C$, ionic species are hydronium ions (H^+), perfluorosulfonic acid (PFSA) membranes as solid electrolytes;
- Solid oxides electrolysis cell (SOEC): working cell temperature $>700°C$, ionic species are oxide ions (O^{2-}), yttrium-stabilized zirconia as solid electrolytes.

Each system involves different reactions at both electrodes according to the ionic species transported through the electrolyte. Eqs. (21)–(26) give the different reactions occurring in the different electrolysis systems:

In an AEC:

$$\text{Anodic reaction: } 4OH^- \rightarrow 2H_2O + 4e^- + O_2 \qquad (21)$$

Hydrogen Electrochemical Production. DOI: http://dx.doi.org/10.1016/B978-0-12-811250-2.00003-0

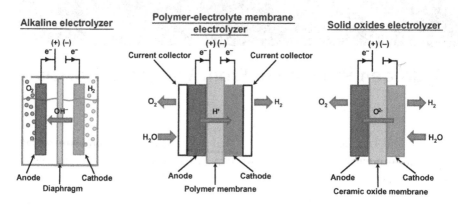

Figure 7 Working principle of the different water electrolysis cell systems.

$$\text{Cathodic reaction: } 2H_2O + 4e^- \rightarrow 4OH^- + 2H_2 \qquad (22)$$

In a PEMEC:

$$\text{Anodic reaction: } 2H_2O \rightarrow 4H^+ + 4e^- + O_2 \qquad (23)$$

$$\text{Cathodic reaction: } 4H^+ + 4e^- \rightarrow 2H_2 \qquad (24)$$

In a SOEC:

$$\text{Anodic reaction: } 2O^{2-} \rightarrow 4e^- + O_2 \qquad (25)$$

$$\text{Cathodic reaction: } 2H_2O + 4e^- \rightarrow 2O^{2-} + 2H_2 \qquad (26)$$

Independently on the electrolytic media, the standard anode potential $\left(E^0_{(O_2/H_2O)}\right)$ for the anodic reaction of water oxidation into dioxygen is

$$E^0_{(O_2/H_2O)} = 1.23 \text{ V } vs \text{ SHE} \qquad (27)$$

while the standard cathode potential $\left(E^0_{(H_2O/H_2)}\right)$ for all cathodic reactions of water reduction into dihydrogen is

$$E^0_{(H_2O/H_2)} = 0.0 \text{ V } vs \text{ SHE} \qquad (28)$$

The standard cell voltage for the global reaction of water dissociation presented in Eq. (20) is 1.23 V independently on the electrolysis system. Fig. 7 gives the working principle of the three main systems.

3.1 THERMODYNAMICS OF WATER ELECTROLYSIS

The standard molar enthalpy of water decomposition, $\Delta_r H$, is the total energy required to split 1 mole of water molecule into 0.5 mole of dioxygen and 1 mole of dihydrogen. A part of this energy corresponds to the thermal energy necessary for the reaction to take place; increasing the thermal energy provided to the system allows reducing the electrical energy required for the reaction of water splitting. Thermodynamic relation is given in Eq. (29):

$$\Delta_r H = \Delta_r G - T \Delta_r S \tag{29}$$

with $\Delta_r G$ the molar Gibbs energy of water decomposition and $\Delta_r S$ the molar entropy of the water splitting reaction.

The Gibbs energy represents the minimum electric energy and $T \Delta_r S$ the minimum heat required for the reaction of water splitting to take place. The electric energy ($\Delta_r G$) will be provided by an external electric generator, and the heat energy ($T \Delta_r S$) will be provided by the working temperature conditions. From Eq. (29), two electrolysis voltages can be defined. The first one, from the Gibbs energy, is to the thermodynamic voltage (U_{Rev}) also called the reversible voltage; the second voltage is the enthalpic voltage (U_{Therm}) more commonly called thermoneutral voltage of the water decomposition reaction. This last voltage represents the global energy requires for the reaction to occur. The reversible and thermoneutral voltages for the water splitting reaction are calculated from Eqs. (30) and (31), respectively:

$$U_{Rev} = \frac{\Delta_r G}{nF} \tag{30}$$

$$U_{Therm} = \frac{\Delta_r H}{nF} \tag{31}$$

where F is the Faraday constant (96,485 C mol^{-1}), and n is the number of electrons exchanged ($n = 2$). $\Delta_r G$ and $\Delta_r H$ values are dependent on the pressure and the temperature of the system. Under standard conditions ($T = 298$ K and $p = 1$ bar $= 10^5$ Pa), water is under liquid phase, whereas oxygen and hydrogen are under gaseous phases. These conditions are often employed for alkaline and acidic electrolysis systems. In these cases, the standard energy values are

$$\Delta_r G^\circ = 237.22 \text{ kJ mol}^{-1} \rightarrow U_{Rev} = \frac{\Delta_r G^\circ}{2F} \approx 1.23 \text{ V} \tag{32}$$

$$\Delta_r H^{\circ} = 285.8 \text{ kJ mol}^{-1} \rightarrow U_{\text{Therm}} = \frac{\Delta_r H^{\circ}}{2F} \approx 1.48 \text{ V} \qquad (33)$$

A supplementary voltage ($U_{\text{Ent}} = 0.25$ V) could be defined, derived from the entropy $\Delta_r S$ ($163.15 \text{ J mol}^{-1} \text{ K}^{-1}$) change, i.e., the heat demand for the reaction to occur. It corresponds to the minimum overvoltage with respect to the reversible voltage to be applied to the electrolysis cell in order to start the water decomposition reaction.

All thermodynamic values ($\Delta_r G$, $\Delta_r H$, $\Delta_r G$, U_{Rev}, and U_{Therm}) are dependent on the temperature. The changes of the values of the state functions, as well as in the heat energy demand ($\Delta_r S$), as a function of the temperature at constant standard pressure are shown in Fig. 8A, whereas those of both reversible and thermoneutral voltage values are shown in Fig. 8B.

The sudden drops or dramatic changes in the slopes occurring at 100°C for all state functions and voltages are related to the change of state of water, i.e., to the vaporization energy. Over both temperature ranges, below and above 100°C, the total energy demand $\Delta_r H$ and consequently the thermoneutral voltage variations are small. On the other hand, the electricity demand and the reversible voltage decrease with increasing the temperature. At last, the heat demand increases.

When the electrolysis cell is running, i.e., when a current flow crosses the cell, an internal resistance appears, which generates a significant quantity of heat due to electrode reactions and energy dispersion by "Joule effect." The heat generated by "Joule effect" helps to promote the water splitting reaction. According to Fig. 8, increasing the working temperature of electrolysis should favor the reaction of water splitting since the electrical demand decreases. At room temperature, the reversible voltage represents 85% of the total energy required for the reaction, whereas at 500°C the rate decreases to 75%. The thermodynamic data for the water splitting reaction and the relationships of thermodynamic state function values with the reversible and thermoneutral voltages lead to three configurations according to the electrolysis cell voltage (U_{cell}):

- $U_{\text{Cell}} < U_{\text{Rev}}$: the reaction does not occur;
- $U_{\text{Rev}} < U_{\text{Cell}} < U_{\text{Therm}}$: extra heat is required to operate the reaction;
- $U_{\text{Therm}} < U_{\text{Cell}}$: the reaction occurs and evolves heat.

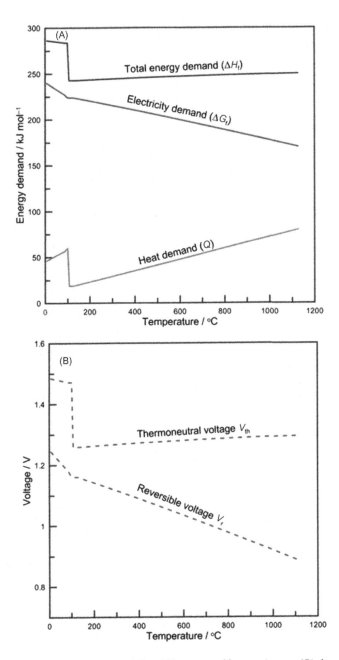

Figure 8 Temperature dependence of (A) enthalpy, Gibbs energy and heat requirement; (B) thermoneutral and reversible voltages for the water splitting reaction in liquid and gaseous phase at a pressure of 1 atm. From A. Goñi-Urtiaga, D. Presvytes, K. Scott, Solid acids as electrolyte materials for proton exchange membrane (PEM) electrolysis: review, Int. J. Hydrogen Energy 37 (2012) 3358–3372 [35].

The reversible potentials of both electrodes, anode and cathode, can be determined using the Nernst equation, which depends on the standard potential, the temperature and the activity of the different chemical species involved in the semielectrochemical reactions. In alkaline media (AEC), the thermodynamic potentials are expressed as follows:

Anode reaction: $E_{Anodic} = E_{(O_2/H_2O)}$

$$= E^0_{(O_2/H_2O)} + \frac{RT}{nF} \ln\left(\frac{(a_{H_2O})^2 (a_{O_2})}{(a_{OH^-})^4}\right) \quad (34)$$

Cathode reaction: $E_{Cathodic} = E_{(H_2O/H_2)} = E^0_{(H_2O/H_2)}$

$$+ \frac{RT}{nF} \ln\left(\frac{(a_{H_2O})^4}{(a_{OH^-})^4 (a_{H_2})^2}\right) \quad (35)$$

where R is the perfect gas constant (8.314 J mol^{-1} K^{-1}), T is the temperature in Kelvin, F is the Faraday constant (96,485 C mol^{-1}), a_x are the activities of the chemical species, and n is the number of electrons involved in the redox reactions. In this case, $n = 4$ since two water molecules are involved.

In acidic media (PEMEC), the thermodynamic potentials are expressed as follows:

Anode: $E_{Anodic} = E_{(O_2/H_2O)} = E^0_{(O_2/H_2O)} + \frac{RT}{nF} \ln\left(\frac{(a_{H^+})^4 (a_{O_2})}{(a_{H_2O})^2}\right) \quad (36)$

Cathode: $E_{Cathodic} = E_{(H_2O/H_2)} = E^0_{(H_2O/H_2)} + \frac{RT}{nF} \ln\left(\frac{(a_{H^+})^4}{(a_{H_2})^2}\right) \quad (37)$

In solid oxides media (SOEC), the thermodynamic potentials are expressed as follows:

Anode: $E_{Anodic} = E_{(O_2/H_2O)} = E^0_{(O_2/H_2O)} + \frac{RT}{nF} \ln\left(\frac{(a_{O_2})}{(a_{O^{2-}})^2}\right) \quad (38)$

Cathode: $E_{Cathodic} = E_{(H_2O/H_2)} = E^0_{(H_2O/H_2)} + \frac{RT}{nF} \ln\left(\frac{(a_{H_2O})^2}{(a_{O^{2-}})^2 (a_{H_2})^2}\right)$

$$(39)$$

The thermodynamic voltage is defined as the potential difference (ΔE) between the anode and the cathode when no current is flowing through the electrolysis cell, i.e., when no reaction occurs at the macroscopic level and is therefore also called the equilibrium cell voltage (U_{Cell}^{Eq}). As it corresponds to the open circuit voltage, it is also referred to as to the "electromotive force." The value of U_{Cell}^{Eq} can be calculated using Eq. (40):

$$U_{Cell}^{Eq} = \Delta E = E_{Anodic} - E_{Cathodic}$$

$$= E_{(O_2/H_2O)}^0 - E_{(H_2O/H_2)}^0 + \frac{RT}{4F} \ln\left(\frac{(a_{H_2})^2 (a_{O_2})}{(a_{H_2O})^2}\right) \tag{40}$$

3.2 KINETIC ASPECTS OF WATER ELECTROLYSIS

The thermodynamic treatment leads to determine the minimum cell voltage to be applied to an electrolysis cell to initiate the water splitting reaction, as a function of temperature and activity of species. But the objective is to produce hydrogen (and oxygen), i.e., to flow an electric current through the electrolysis cell. The reaction kinetics at the electrodes are not infinite, and these limitations involves the appearance of oxidation and reduction overpotentials, $\eta_{Anodic}(i)$ and $\eta_{Cathodic}(i)$, respectively, leading to an overvoltage, i.e., a difference between the voltage applied to the cell during operation and the value of the reversible potential for hydrogen and oxygen production. Moreover, the different elements constituting an electrolysis cell and the interfaces between them (current collectors, connections, interfaces, electrolyte materials, anodic and cathodic catalytic layers) are responsible of the appearance of an ohmic resistance (R_{Cell}). As a consequence, the cell voltage (U_{Cell}) has to be higher than the reversible one for the significant production of hydrogen at the cathode. Taking into account these limitations, the overall cell voltage can be expressed as follows:

$$U_{Cell} = U_{Cell}^{Eq} + \eta_{Anodic}(i) + \eta_{Cathodic}(i) + iR_{Cell} \tag{41}$$

iR_{Cell} being called the ohmic drop.

Both the overpotentials and the ohmic drop vary with the current density applied to the electrolysis cell, as shown in Fig. 9.

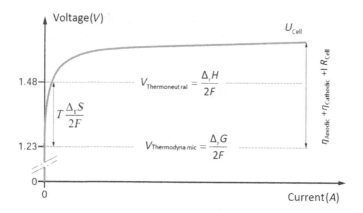

Figure 9 Change in the cell voltage as a function of the applied current in the case of water electrolysis in acidic media at 25° C and atmospheric pressure. From P. Millet, Électrolyseurs de l'eau à membrane acide, Techniques de l'ingénieur, 2007 [36].

In electrochemistry, the rate of a redox reaction is generally controlled by the electron transfer process.

$$Ox + ne^- \leftrightarrow Red \qquad (42)$$

with Ox the oxidizing species, Red the reducing species and n the number of electrons transferred for the redox reaction. The Butler–Volmer relationship allows determining the theoretical currents density (j) for a specific reaction and is written as follows (the current density being the current divided by the electrode surface):

$$j = j^0 \left(\frac{[Red]_{El}}{[Red]_{Sol}} \exp\left(\frac{\alpha nF}{RT} \eta_{Anodic} \right) - \frac{[Ox]_{El}}{[Ox]_{Sol}} \exp\left(\frac{\beta nF}{RT} \eta_{Cathodic} \right) \right) \qquad (43)$$

where $[Red]_{El}$, $[Red]_{Sol}$, $[Ox]_{El}$, and $[Ox]_{Sol}$ are the reducing and oxidizing species at the electrode interfaces and in the bulk solution, respectively, j^0 is the exchange current density, α and β are the anodic and cathodic transfer coefficients ($\alpha + \beta = 1$). In a first approximation, considering that there is no concentration gradient between the bulk solution and the electrode surfaces (no diffusion limitation), the Butler–Volmer relationship can be simplified as follows:

$$j = j^0 \left(\exp\left(\frac{\alpha nF}{RT} \eta_{Anodic} \right) - \exp\left(\frac{\beta nF}{RT} \eta_{Cathodic} \right) \right) \qquad (44)$$

In a second approximation, considering that the anodic overvoltage is large enough (e.g., in the case of water oxidation), the cathodic component becomes negligible with respect to the anodic one and the relationship becomes

$$j = j^0 \exp\left(\frac{\alpha nF}{RT}\eta_{Anodic}\right) \qquad (45)$$

while for the study of an electrochemical cathodic process, if $\eta_{Cathodic}$ is large enough, then the anodic component becomes negligible. The relationship is simplified as

$$j = -j^0 \exp\left(\frac{\beta nF}{RT}\eta_{Cathodic}\right) \qquad (46)$$

From both these equations, the Tafel equations can be expressed as the over-voltages as a function of the current density:

$$\eta_{Anodic} = a + b\,\log j; \text{ with } a = \frac{-2.3RT}{\alpha nF}\log(j^0) \text{ and } b = \frac{2.3RT}{\alpha nF} \qquad (47)$$

and:

$$\eta_{Cathodic} = a' + b'\log j; \text{ with } a' = \frac{2.3RT}{\beta nF}\log(j^0) \text{ and } b' = \frac{-2.3RT}{\beta nF} \qquad (48)$$

Based on these equations, the exchange current density (j^0), the transfer coefficients (α and β), and the Tafel slope values (a and a') can be determined from experimental polarization curves. These kinetic data allow proposing the rate determining step (rds) of the reactions at the electrode material. They are dependent on the electrocatalytic material, and they can serve to compare the catalytic activities and efficiencies of the catalysts.

As previously explained, the cell voltage (U_{Cell}) under working conditions (when current density is applied to the electrolysis cell) is higher than the equilibrium value (U_{Cell}^{Eq}) due to both the irreversibility of the electrochemical reactions (presence of overpotentials) and the cell resistance. So the overall energy efficiency ε_{Cell} (neglecting the energy consumed in the auxiliary equipment) can be calculated as follows [37]:

$$\varepsilon_{Cell} = \frac{\Delta_r H^{Rev}}{\Delta_r G^{Rev} + T\Delta_r S^{Rev} + nF(|\eta_{Anodic}| + |\eta_{Cathodic}| + jR_{Cell})} \qquad (49)$$

where $\Delta_r H^{Rev}$, $\Delta_r G^{Rev}$, and $\Delta_r S^{Rev}$ are the enthalpy, the Gibbs energy, and the entropy at the equilibrium ($j = 0$).

Under low-temperature working conditions (PEM and alkaline electrolyzer), a simplified equation can be used assuming that the heat demand is small compared to the global energy of the reaction. The energy efficiency can be expressed as a function of U_{Cell}^{Eq} and U_{Cell}:

$$\varepsilon_{Cell} = \frac{\Delta_r G^{Rev} + T\Delta_r S^{Rev}}{\Delta_r G^{Rev} + T\Delta_r S^{Rev} + nF(|\eta_{Anodic}| + |\eta_{Cathodic}| + jR_{Cell})} \quad (50)$$

with $\Delta_r G^{rev} = nFU_{Cell}^{Eq}$ and $U_{Cell} = U_{Cell}^{Eq} + |\eta_{Anodic}| + |\eta_{Cathodic}| + jR_{Cell}$ then,

$$\varepsilon_{Cell} = \frac{U_{Cell}^{Rev} + \left(T\Delta_r S^{Rev}/nF\right)}{U_{Cell} + \left(T\Delta_r S^{Rev}/nF\right)}$$

$$= \frac{U_{Cell}^{Rev}}{U_{Cell}} \left(\frac{1 + T\Delta_r S^{Rev}/\Delta_r G^{Rev}}{1 + T\Delta_r S^{Rev}/(\Delta_r G^{Rev} + nF(|\eta_{Anodic}| + |\eta_{Cathodic}| + jR_{Cell}))} \right)$$

$$(51)$$

and $\varepsilon_{Cell} \approx \dfrac{U_{Cell}^{Rev}}{U_{Cell}}$ since it is considered that $\left(\dfrac{T\Delta_r S^{Rev}}{\Delta_r G^{Rev}}\right) \ll 1$.

3.3 ACIDIC WATER ELECTROLYSIS

Water electrolysis processes are essentially constrained by the thermo-dynamically unfavorable oxygen evolution reaction (OER). Therefore, this section will mainly focus on electrocatalytic anodic materials.

3.3.1 Electrodes Materials

From voltammetric measurements, Miles et al. [38,39] determined the over-potential of the OER and hydrogen evolution reaction (HER) for different pure and nonsupported metals in acidic medium (Fig. 10A). The order of activity for the OER was Ru ≈ Ir > Pd > Rh > Pt > Au > Nb, while for the HER it was Pd > Pt ≈ Rh > Ir > Re > Os ≈ Ru > Ni. Many catalytic materials were then developed on the basis of these metallic elements for the elaboration of anodic and cathodic electrodes.

However, for the OER, it is well known that metal oxides are more stable than pure metals. Trasatti has then compared the activity in terms of overpotential of different metal oxides for the OER and found

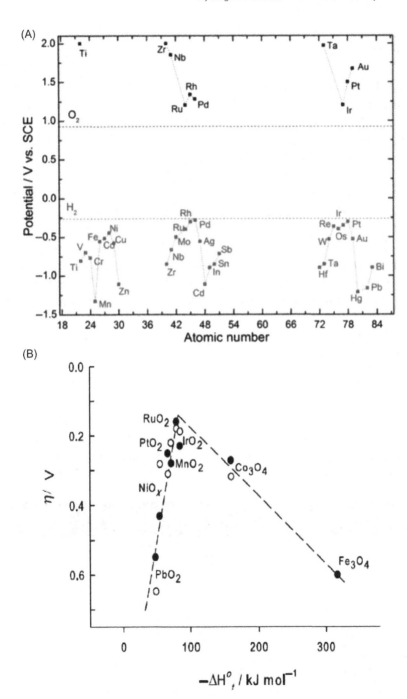

Figure 10 (A) Onset potential for the OER and HER for different metallic elements (potentials determined at 2 mA cm⁻² by cyclic voltammetry measurements at 50 mV s⁻¹ in 0.1 mol L⁻¹ H₂SO₄ at 80°C) [38]; (B) electro-catalytic activity towards the OER of various oxides as a function of the enthalpy of the lower → higher oxide transition, in acid (•) and alkaline (∘) solutions [40].

that the most active were RuO_2 and IrO_2, in both alkaline and acidic electrolytes (Fig. 10B).

3.3.1.1 Anodic Electrode Materials

Trasatti and coworkers [40−42] enounced three main reasons explaining the difficulty to understand the OER mechanism occurring on catalytic materials: (1) the high activation energies of intermediates formed during the reaction involve complex reaction pathways that are sensitive to electrode surface properties, (2) high anode potentials are required for the OER reaction that may lead to change in the structure of the electrode surface, and (3) the surface changes of the electrodes under potential control lead to modify the kinetic of the reaction with time. The composition and structure of the electrocatalysts play an important role for the OER and particularly determine the mechanism occurring on their surface. Hence, the synthesis methods should confer the most favorable structure to the electrocatalysts for achieving the highest OER efficiency as possible. Trasatti et al. [40] have summarized the main requirements for technological applications of electrodes or electrocatalysts and the main structural parameters influencing the OER efficiency (Table 2).

In the OER potential range, metal corrosion and passivation were observed. Taking into account the high anode potentials for OER, the anodic materials will be under oxide forms, at least on their surface. These oxide species have to display good electrical conductivity and to be stable to avoid further oxidation processes. The OER reaction

Table 2 Main Requirements for Electrocatalytic Materials and Factors Influencing the OER Reaction [40,43]	
Main Requirements	Influencing Factors
High surface area	Chemical nature of the catalyst
High electrical conduction	Morphology and microstructure
Good electrocatalytic properties	Nonstoichiometry (ionic and electronic defects, etc.)
Long-term mechanical and chemical stabilities	Magnetic properties
Low gas bubble problems	Band structure of the oxide/Bond strength of M−O
Enhanced selectivity	Number of electrons in d band
Availability and low cost of materials	Effective Bohr magneton
Health safety	Surface electronic structure
	Crystal-field stabilization energy
	Synergetic effects (mixed or doped oxides)

involves formation and rupture of bonds between the metal oxide, the water molecules, and the oxygen atoms. The main steps of the reaction are the adsorption of water molecules forming intermediate species, reaction between adsorbed species and oxygen desorption, involving changes in metal oxidation state. Trasatti et al. [40] showed that materials with intermediate values of the enthalpy for the transition of a metal oxide to higher oxidation state, ΔH_t^0, such as RuO_2 and IrO_2, led to higher activity in terms of lower overpotentials for the OER (Fig. 10B).

In addition, according to Matsumoto et al. [43], the rate of desorption and/or adsorption of the reactive species is depending on the strength of the metal−oxygen bond of the intermediate reaction species; the stronger is the bond strength, the lower is the reaction kinetics. At last, the electron-transfer kinetics, which are dependent on the Fermi level, also play an important role in the catalytic activity of oxide materials. Moreover, Rogers et al. [44] have shown that the electronic transport in the oxide lattice is correlated to the crystallographic structure. For example, the rutile structures of RuO_2 and IrO_2 are semiconductive.

Based on these previous works, many studies were performed on ruthenium and iridium oxides owing to their high activities towards the OER. RuO_2 is the most efficient material for the OER but presents a poor stability and can be further oxidized into RuO_4 [45−48], whereas IrO_2 is slightly less active but is more stable [49]. Thereby, mixed oxides were elaborated with the hope to combine the properties of each pure oxide material:

• Bimetallic oxides composed of RuO_2 and IrO_2 with different ratios,
• Mixed oxides composed of RuO_2 or IrO_2 and other non-noble metal oxides (MO_2 with M = Sn, Ti, Mn, Ce, Ta, Nb).

In all cases, RuO_2 and IrO_2 are the materials conferring the electrocatalytic properties for the OER, whereas the non-noble oxides help to increase the stability of the electrodes and to decrease the amount of noble metals (Ru and Ir). In order to decrease the metal amount in electrodes, recent works were devoted to the synthesis of nanoscopic catalysts on a high surface area electron conductive support with high corrosion resistance, such as Sb-doped SnO_2 (ATO for antimony tin oxide) [50,51], titanium suboxides (TiO_x) [52], titanium carbides (TiC) [53], titanium nitrides (TiN) [54], titanium carbonitrides (TiCN) [55], etc.

Table 3 Most Common Characterization Methods for Electrocatalytic Oxides and Information Provided	
Method	Property/Information Obtained
Temperature programed analysis (TPA)	Temperature of decomposition, precursor/intermediate interactions
X-ray diffraction (XRD)	Crystal structure, crystallinity, crystal size
BET surface area	Real surface area
Electrical resistance	Electronic structure
X-ray photoelectron spectroscopy (XPS)	Surface composition electronic interactions
X-ray absorption spectroscopy (XANES, EXAFS)	Valance, atomic structure structure-potential relationship
Transmission electron microscopy (TEM) coupled with electron dispersive spectroscopy (EDS)	Morphological structure, particle size and bulk atomic composition

3.3.1.1.1 Physicochemical Analysis

As the high efficiency of catalysts towards the OER is dependent on their physicochemical properties (composition, nature, mean size of the particles and crystallites, etc.), many physicochemical methods were used for their characterization. Trasatti et al. [56] have summarized the main physicochemical methods of characterization and the important information they provide concerning the catalytic materials (Table 3).

The contribution of some characterization to the understanding of the properties of catalytic materials will be succinctly presented, as examples. Transmission electron microscopy (TEM)/electron dispersive spectroscopy (EDS) measurements allow determining the morphology of the oxide materials, the particle size, their size distribution, the formation of agglomerates, and the bulk composition (Fig. 11A). The activity towards OER being dependent on the metal oxide structure, X-ray diffraction (XRD) analyses were realized to determine the crystal size, the crystallinity, and the crystal lattice parameters of the materials. Characteristic diffractograms of rutile $Ru_{(1-x)}Ir_xO_2$ materials are presented in Fig. 11B. The narrower are the diffraction peaks, the higher is the crystallinity of the materials, giving information on the crystallite size. Raman spectroscopy can also be used to complete or confirm the nature and structure of the synthesized particles (Fig. 11C). X-ray photoelectron spectroscopy (XPS) analysis is useful to determine the surface elemental composition of the oxide materials, their oxidation level, and their electronic interactions (Fig. 11D). In

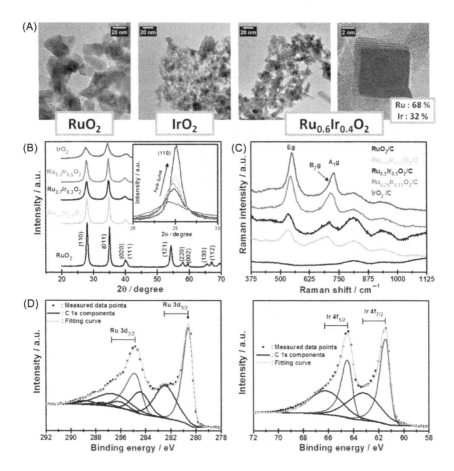

Figure 11 (A) TEM images of mono and bimetallic oxides electrocatalysts (modified from [74]), (B) XRD patterns of mono and bimetallic oxides electrocatalysts (modified from [74]), (C) Raman spectra of $Ru_xIr_{(1-x)}O_2/C$ catalysts obtained at 4 mW (modified from [75]), (D) XPS spectra of the Ru 3d (left) and Ir 4f (right) for $IrO_2@RuO_2$ catalyst. Modified from T. Audichon, T.W. Napporn, C. Canaff, C. Morais, C. Comminges, K.B. Kokoh, IrO2 coated on RuO2 as efficient and stable electroactive nanocatalysts for electrochemical water splitting, J. Phys. Chem. C. 120 (2016) 2562–2573 [76].

the case of mixed oxides, these methods can allow evidencing a surface enrichment or impoverishment of a given metal.

Cruz et al. [57] have synthesized RuO_2 oxide material by a colloidal method followed by a calcination step between 200 and 350°C. At low calcination temperatures, hydrated ruthenium oxides are obtained, whereas for high calcination temperatures the crystallinity of RuO_2 particles increased. Same behaviors were reported by Zheng et al. [58] from a sol–gel synthesis method and by Devadas et al. [59] from the

"Instant Method" assisted by microwave irradiation. Tsuji et al. [60] demonstrated that for temperatures below 200°C, materials were amorphous during their elaboration by electrodeposition or by sputtering. The materials become crystalline only for calcination temperatures above 300°C. These results were confirmed by Sassoye et al. [61] for metal oxides prepared by a colloidal synthesis method in aqueous phase. However, according to Kim et al. [62], the presence of crystalline RuO_2 has been detected by Raman spectra from 100°C although the XRD patterns did not present any diffraction peak for material calcined at low temperature (below 200°C). The high level of RuO_2 hydration at low calcination temperatures was proposed to explain the absence of XRD diffraction peaks. Audichon et al. [75] obtained same results for $Ru_{(1-x)}Ir_xO_2$ mixed oxides. In addition, Murakami et al. [63] showed that the crystallite size of metal oxides increased with the calcination temperature, whereas in parallel the specific surface area of the oxide materials decreased. As a general trend, the oxides become crystalline for calcination temperatures between 300°C and 400°C, and the particle sizes reach a few tens of nanometers [64], although the metal oxide microstructure and the particle shape can slightly differ as a function of the synthesis route.

For mixed oxide materials such as $Ir_xRu_{(1-x)}O_2$, $Ir_xSn_{(1-x)}O_2$, $Ru_xSn_{(1-x)}O_2$, $Ru_xMn_{(1-x)}O_2$, and $IrO_2-Ta_2O_5$ [65–69], with x varying from 0 to 1, the determination of the atomic ratio from XRD measurements is made difficult because all oxides lead the rutile crystallographic structure, and all metals have very close atomic radii. Therefore, the locations of the different diffraction peaks are very close and can hardly be separated. But Owe et al. [70] performed high-resolution XRD measurements on $Ir_{(1-x)}Ru_xO_2$ materials and observed a gradual shift of the high Miller index diffraction peaks with the oxide composition, confirming the formation of a single-phase solid solution. Audichon et al. [74] also observed a peak shift and concluded that a homogeneous mixed oxide phase was obtained (Fig. 6B). On the other hand, using a "Pechini" method and a direct, dry, jet-flame-based process, Mamaca et al. [71] and Roller et al. [72], respectively, have observed the presence of two phases on their XRD patterns—the first one being attributed to the oxide materials with the rutile structure and the second one to a metallic phase. But most of the syntheses allowed obtaining a homogeneous mixed oxide phase [67,73], as often confirmed by TEM−EDS measurements [52,74].

TEM measurements have been performed to determine the morphology of mixed oxides. No important change was observed when comparing the materials obtained by different synthesis methods;, the particle sizes were always comprised between 20 and 100 nm, they always displayed cuboic shapes and they were generally agglomerated. [67,77,78]. Corona-Guinto et al. [79] used an EDS mapping method to verify the homogeneity in the repartition of the metals in a RuIrCoO$_x$ electrocatalyst. The effect of the addition of a second metal oxide in the catalyst composition was studied by XRD. As for pure oxides, Baglio et al. [80] showed that for the same oxide composition, increasing the calcination temperature led to an increase of the mean crystallite size; these authors used X-ray fluorescence spectroscopy to quantify the oxides present in the catalytic materials. Brunauer, Emmett and Teller theory (BET) analyses were also used to determinate the effect of the addition of a second metal oxide on the specific surface area [81]. Dependently on the crystallization temperature of both oxides, the BET specific surface area values tend to increase or to decrease with the addition of the second metal oxide: if the added metal oxide had a lower crystallization temperature than the catalytic metal oxide, the surface area increased; and if the added metal oxide had a higher crystallization temperature, the surface area decreased [74]. XPS has mainly been used to determine the nature of the oxides, the metal oxidation states and the surface composition as a function of the synthesis method, but this characterization method has also allowed observing the evolution of these parameters during electrochemical measurements [46]. RuO$_2$ and IrO$_2$ metal oxides give XPS core level spectra with characteristic shapes presenting a doublet of asymmetric peaks for Ru 3d and Ir 4f [82,83] that are attributed to the + IV oxidation level of ruthenium and iridium. The fitting of XPS experimental spectra takes generally two components into account [82,84,85]. The first predominant peak doublet attributed to the primary spin−orbit components is correlated to the oxide structure. The second component, accompanied by two satellite peaks close to the first components, is attributed to the final state screening effects. However, the reason and the nature of these screening effects are not clearly established; the satellite peaks are also related to Ru^{4+} and Ir^{4+} but with slightly different environments, likely hydroxyl or oxohydroxyl groups [82]. The O 1s core level spectra do not allow determining clearly the nature of the oxygenated groups due to sample contamination [75]. Sometimes, the fitting procedure is applied on the Ru 3p core level spectrum because the C 1s contribution linked to surface sample contamination appears in the same binding energy range than Ru 3d [86]. From XPS measurements, Audichon et al. [76] showed that in the case of IrO_2 coated on RuO_2 materials, the Ir ratio was higher on the surface than in the core of the catalytic particles (Fig. 11).

At last, the conductivity of oxide materials is also a very important parameter to evaluate their catalytic activity. The study of Huang et al. [87] on the effect of the calcination temperature for the preparation of RuO_2 thin film by a metal–organic chemical vapor deposition (MOCVD) method demonstrated that for temperatures higher than 100°C, the electrical properties of materials displayed metallic character: The resistivity of the thin film decreased when increasing the calcination temperature. The resistivity also decreased when increasing the thickness of the metal oxide film. Barbieri et al. [88] showed that the electronic conductivity of dry materials increased when increasing the calcination temperature of the RuO_2 nanoparticles. In hydrated oxides, the water content in the crystal lattice acted as a barrier for the electrical conduction, whereas the decrease of water content in the oxide lattice after heat treatment increased the conductivity.

3.3.1.1.2 Electrochemical Behavior
RuO_2 and IrO_2 are the most studied metal oxides owing to their high activities, whereas SnO_2 and TiO_2 are often added owing to their good stability in acidic media and also to decrease the noble metal loading in the catalyst composition. Electrochemical measurements are conducted to correlate the activity to the structural properties of the electrocatalytic materials, particularly by evaluating the specific capacitances, the electric charges, the kinetics for the OER (exchange current densities, Tafel slopes, charge transfer resistance), and the stability of the materials during long-term experiments.

The first electrochemical study consists in recording cyclic voltammograms (CV) in H_2SO_4 supporting electrolyte in the potential range from 0.05 to 1.4 V *vs* reversible hydrogen electrode (RHE). Oxide materials display characteristic CV according to their composition. For example, the typical CV of crystalline RuO_2 (blue line) in Fig. 12A is different from that of crystalline IrO_2 [red line]; it also differs from that of hydrated RuO_2 (black line). The current densities are due to either capacitive or faradaic contributions at the electrode surface, so they are dependent on the material structure and composition. For example, hydrated ruthenium oxide displays higher current densities than the crystalline one over the potential range studied.

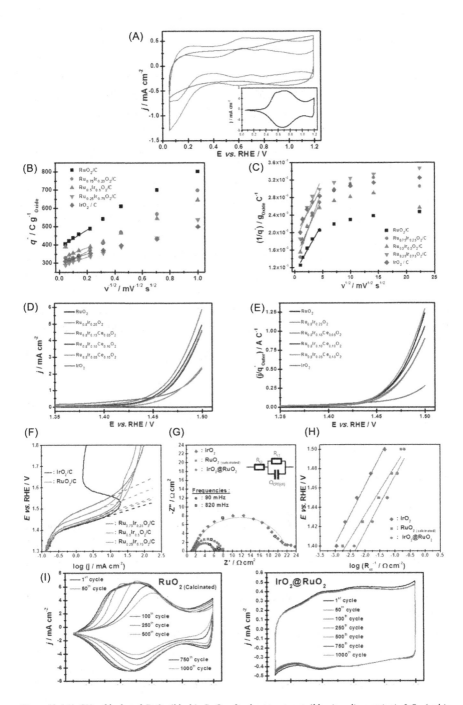

*Figure 12 (A) CVs of hydrated RuO₂ (black), RuO₂ after heat treatment (blue in online version), IrO₂ (red in online version), and IrO₂@RuO₂ catalysts (green in online version) recorded at 20 mV s⁻¹ [76]; (B,C) voltam-metric charges vs scan rate, (B) extrapolation of the most accessible charges (q*ₒᵤₜₑᵣ) and (C) extrapolation of the total charges (q*Total) [75]; (D) OER polarization curves on RuO₂, IrO₂ and Ru₀.₈Ir₍₀.₂₋ₓ₎CeₓO₂ electrodes normalized with the electrode surface area and (E) with the most accessible charges (q*ₒᵤₜₑᵣ) [52]; (F) Tafel plots for RuₓIr₍₁₋ₓ₎O₂/C [75]; (G) Nyquist plots for mono- and coated oxide electrodes recorded at 1.5 V vs RHE during the OER [76]; (H) Tafel plots based on EIS data [76]; (I) CV during repetitive potential cycles [88]. All measurements were recorded in 0.5 mol L⁻¹ H₂SO₄ electrolyte and 25°C.*

The capacitance of the materials can be evaluated from the double-layer potential region of the CV. The larger is this double-layer region, the closer to the electrocatalyst active sites are the electroactive species favorable to OER. Barbieri et al. [88] evaluated the capacitance values of RuO_2 materials as a function of the heat treatment temperature. The capacitance values are invariant for low heat treatment temperatures, start to decrease, and tend to stabilize for temperatures higher than 200°C. The change in capacitance has been directly correlated to the hydration level of the oxide materials: the higher the hydration level, the higher the capacitance value. But, for a too high hydration level, the accumulation of water molecules can act as an electrical barrier decreasing the capacitance values, as observed by Barbieri et al. [88] at 25°C. For Trasatti et al. [89], the capacitive properties of metal oxides were not only related to the double-layer capacity, but also to the faradaic reactions occurring at the interface oxide/electrolyte:

$$RuO_2 + \delta H^+ + \delta e^- \leftrightarrow RuO_{2-\delta}(OH)_\delta \tag{52}$$

This equation has been generalized to other metal oxides than RuO_2 [75,88] and considers that in aqueous solution, the metal oxide surface is hydrated to form hydroxide species:

$$MO_x(OH)_y + \delta H^+ + \delta e^- \leftrightarrow MO_{x-\delta}(OH)_{y+\delta} \tag{53}$$

where M is the metal species.

The faradaic reactions give rise to oxidation and reduction peaks on the CV attributed to the Ru(III)/Ru(IV), Ru(IV/VI) [90], and Ru(VI)/Ru(VIII) [91] redox transitions for RuO_2, and Ir(III)/Ir(IV), Ir(IV)/Ir(VI) [91] for IrO_2. For mixed oxides, these contributions overlap on the potential range studied and the presence of two metals can slightly shift the potential of each redox transition [74]. Furthermore, the addition of other active metal oxides can lead to supplementary peaks attributed to their redox transition [52]. Due to the contribution of capacitive and faradaic processes over the same potential range, the term of pseudocapacitance can be used. The specific capacitance or pseudocapacitance values are determined by integration of the CVs curves, according to the following equation:

$$C\left(F\ g^{-1}\right) = \frac{1}{vm(E_2 - E_1)} \int_{E_1}^{E_2} i(E)dE \tag{54}$$

where v is the scan rate ($V\,s^{-1}$), m is the mass of oxide material deposited on the working electrode in mg, E_1 and E_2 are the lower and upper potential limits and $i(E)$ is the current at electrode potential E.

CV on Fig. 12A shows a faradaic cathodic current feature between 0.05 and 0.3 V vs RHE for crystalline RuO_2 and IrO_2 materials. This nonreversible peak was attributed to the insertion or absorption of hydrogen in the oxide lattice and directly linked to the sample crystallinity. Studies of CV recorded with different potential limits performed by Juodkazis et al. [92] and Audichon et al. [93] have revealed that the reduction products are oxidized over the whole anodic scan. This contribution can interfere with the capacitance measurements. Sugimoto et al. [94,95] carried out a complete study of the capacitance values as a function of the potential range and scan rate. When increasing the scan rate, the capacitance decreases. At high scan rates only the most accessible oxide sites in the catalytic layer are contributing to the double-layer capacity, whereas at low scan rates the whole active sites are contributing. This phenomenon is correlated to the diffusion of the electroactives species from the electrolyte to the catalytic sites [96]. Sugimoto et al. [95] proposed that three contributions to the capacitance have to be considered: (1) the electric double-layer capacitance (C_{dl}) that is constant independently on the scan rates, (2) the capacitance induced by the electrosorption of ionic species on the oxide surface (C_{ad}), which decreases when increasing the scan rate value, and (3) the capacitance induced by electrochemical irreversible faradaic reaction (C_{irr}) that decreases when increasing the scan rate. These authors demonstrated that the capacitance decreased with the increase of the metal oxide loading and of the particles size. In addition, the decrease of the hydration level and the increase of the crystallinity and of the crystallite size lead to the decrease the capacitive properties [97]. In the case of mixed oxides, the capacitances depend on the composition owing to the differences in dehydration and crystallization temperatures of the pure oxide species. Best capacitive properties were obtained for oxide materials having both particles of small sizes and good crystallinity.

It is difficult to determine the number of active site in a metal oxide catalytic layer because faradaic, and nonfaradaic contributions are involved over the stability potential range. But it has been proposed that the voltammetric charges (q^*) were proportional to the number of active sites [41,98]. The voltammetric charges can be evaluated from CV measurements according to the following equation:

$$q^*(C) = \frac{1}{v} \int_{E_1}^{E_2} i(E)dE \tag{55}$$

The integration of the CV allows determining the number of active sites involved in capacitive contribution and in faradaic reaction of surface hydroxide formation. The reversibility of these contributions can be evaluated from either the charges or the capacitance values by determining the anodic charge/cathodic charge ratio or anodic capacitance/cathodic capacitance ratio, respectively. The closer to 1 are the values of these ratios, the more reversible is the reaction [90]. Sugimoto et al. showed the nonreversibility of the capacitance [94], which also applies to the charge ratio.

The number of charges decreases with the decrease of the oxide hydration level and with the increase of the heat treatment temperature [99]. The number of charges measured is also dependent on the scan rate: at a very low scan rate, the electroreactive species are allowed to diffuse inside the porosity of the catalytic layer and to react with all the active sites, whereas at high scan rates the reaction can only take place on the most accessible active sites. Based on these diffusion limitations, Ardizonne et al. [98] established two equations to calculate the total charges (q_{Total}^*) and the most accessible charges (q_{outer}^*) when the scan rate values tend to 0 and ∞, respectively.

$$\frac{1}{q^*} = \frac{1}{q_{Total}^*} + C_1\sqrt{v} \tag{56}$$

$$q^* = q_{Outer}^* + C_2\frac{1}{\sqrt{v}} \tag{57}$$

where C_1 and C_2 are constants, v is the scan rate, and q^* is the average charge calculated for each scan rate v.

The q_{outer}^* values are obtained by determining the intercepts of the linear part of the curves for high scan rates (Fig. 12B), whereas the q^*_{Total} values are obtained for low scan rates (Fig. 12C). The active site accessibility (q^*_{Outer}/q^*_{Total}) can then be calculated. The values of the accessibility calculated from the determination of q^*_{Outer}/q^*_{Total} as a function of the composition of mixed oxides do not follow a linear trend [74,100], indicating that the accessibility is dependent on the material composition and probably on the synthesis route. The particle and crystallite sizes and the hydration level of the oxides depend on the heat treatment temperature and atmosphere, on the metal oxides composing the mixed oxides and on their ratios [65]. Results indicated that the best electrocatalytic activities are achieved with materials presenting well dispersed small nanoparticles leading to very high active surface area with a large number of active sites and high accessibility.

The electrocatalytic activities of the oxide materials towards the OER are generally determined by recording polarization curves with upper potential limits higher than 1.5 V *vs* RHE (Fig. 12D). The OER onset potential generally higher than 1.375 V *vs* RHE as well as the slope of increasing current depends on the catalyst composition, so that the electrocatalytic activities of materials are determined by the value of the overpotentials for a given current density in order to avoid the difficulty of the determination of the onset potential. The lower is the overpotential at a given current density, the higher is the catalytic activity of the oxide material towards the OER [74]. The electrocatalytic performances depend on the heat treatment temperature and on the composition of mixed oxides [101]. Siracusano et al. [102] recorded polarization curves at different temperatures on the same electrode material. The OER onset potential decreased with the increase of temperature. Lodi et al. [41] have determined the electrochemical activation energies (E_a) from the Arrhenius plots. They found that the E_a for the OER decreased when the electrode potential increased.

The activity has been defined as the overpotential for a given current. However, the current can be normalized either with respect to the geometric electrode surface area in order to compare the intrinsic activity of the catalysts for a given loading (Fig. 12D) [52,65] or with respect to the charge determined from Eqs. (56) and (57) [70,103] in order to obtain the electrocatalytic activities by active sites (Fig. 12E). By using this last method, Audichon et al. [74] have evidenced beneficial effects in $Ru_xIr_{(1-x)}O_2$ mixed oxides on overpotentials, which were lower than those obtained with the pure oxides.

The Tafel plots can be drawn from the polarization curves (Fig. 12F). The plots present two linear parts (one for low overpotentials and the other for high overpotentials), the slope of which are useful to determine the rate determining step (rds) of the reaction. The first Tafel slope obtained at low overpotentials allows determining the very initial rds of the reaction. Owing to the complex reaction of the OER, involving multielectrons transfer, several mechanisms involving different rate-determining steps were proposed, leading to different Tafel slopes [43]. The generally accepted mechanism starts from a water adsorption step on an active site (S) accompanied with a charge transfer leading to an adsorbed hydroxyl intermediate:

$$S + H_2O \rightarrow S - OH_{ads} + H^+ + e^- \qquad (58)$$

the second step consists in a deprotonation reaction of the adsorbed intermediate leading to a second adsorbed intermediate:

$$S - OH_{ads} \rightarrow S - O_{ads} + H^+ + e^- \qquad (59)$$

at last, reaction between two adsorbed intermediates leads to the formation of O_2:

$$2S - O_{ads} \rightarrow 2S + O_2 \qquad (60)$$

Elsewhere it has been shown that the Tafel slopes are 120, 40, and 15 mV dec^{-1} for reactions (58)–(60) being the rds, respectively [104]. However, other authors [104,105] explained that the bond strength of intermediates to active sites could differ owing to the catalytic layer composition, leading to other Tafel slope values. They proposed that reaction (58) could be separated into two steps:

$$S + H_2O \rightarrow S - OH^*_{ads} + H^+ + e^- \qquad (61)$$

$$S - OH^*_{ads} \rightarrow S - OH_{ads} \qquad (62)$$

where the intermediate species $S - OH^*_{ads}$ and $S - OH_{ads}$ have the same chemical structure but different bond strengths according to the active site nature. Both steps can occur alternatively or in parallel, and the corresponding Tafel slope is 60 mV dec^{-1}. In addition, depending on the bond strength of the intermediate species, a chemical reaction can take place in parallel to step (59):

$$2S - OH_{ads} \rightarrow S - O_{ads} + S + H_2O \qquad (63)$$

The Tafel slope associated with this rds is 30 mV dec^{-1}. Most of studies lead to Tafel slope between 40 and 48 mV dec^{-1} depending on catalyst composition and catalytic layer structure, indicating that the rds for the OER is reaction (59) [52,70,74,100]. On the one hand,

Owe et al. [70] and Kotz et al. [46] demonstrated on the basis of Tafel slope determination that synergetic effect improving the kinetic parameters of the reaction took place in mixed oxide materials. Also, Lodi et al. [41] showed that the morphology, particularly the compactness, of the electrocatalytic layer could lead to higher Tafel slope value. The fitting of the Tafel plots allows also determining the exchange current density (j_0) that represents the intrinsic activity of the catalysts towards the OER. Audichon et al. [93] found exchange current density (j_0) values in the range from 10^{-5} to $10^{-4}\,\mathrm{mA\,cm^{-2}}$ for the OER on $Ru_xIr_{(1-x)}O_2/C$ mixed oxides, showing that the OER is a slow process.

The kinetic studies can be completed by electrochemical impedance spectroscopy (EIS) measurements in a three-electrode cell at different potentials over the capacitive region or over the OER region. The experimental data are fitted with electrical equivalent circuit to determine the values of the double-layer capacitance (C_{dl}), the charge transfer (R_{ct}), and the electrolyte resistance (R_s). Typical Nyquist diagrams obtained in the OER potential region are shown Fig. 12G. The Nyquist diagrams and the Bode plots depend on the potential applied for the measurement and on the catalyst composition and/or the catalytic layer structure/morphology. In many works, two semicircles are observed. The first one, poorly defined at high frequencies, is attributed to the behavior of the porous oxide layer; the second one at lower frequencies is attributed to the OER process. In such cases, the experimental data are modeled with an equivalent electrical circuit (EEC) based on two capacitive loops, represented by $R_s(R_fC_f)(R_{ct}C_{dl})$ [106]. R_s is the cell resistance (including electrolyte and connections), R_f is the resistance of oxide layer, and C_f is the capacitance. The (R_fC_f) combination is independent of the potential. R_{ct} is the charge transfer for the OER, and C_{dl} is the double-layer capacitance. In others works, only one semicircle is observed, and only one capacitive loop is used to model the experimental data [93]. The EEC corresponds to $R_s(R_{ct}C_{dl})$ [106]. In these cases, the catalytic layer contributions are probably diluted inside the large depressed semicircle for the capacitance contribution and in the electrolyte resistance for the electrode material resistance contribution [107]. Then, C_{dl} includes the double-layer capacitance due to the electrode polarization and the capacitance attributed to the oxide thin layer structure. In both case, the fitting procedure shows a better agreement between experimental and modeled data when the capacitance is replaced by a constant phase element (CPE). CPE is often used to model depressed semicircles due to heterogeneity and roughness of the electrode surface. Independently on the considered EEC, the increase of the electrode potential translates into the decrease of the semi-circle related to the OER reaction, indicating a better charge transfer on the oxide material surface [93,103]. Simultaneously, C_{dl} value increases indicating that the charge accumulation at the interface electrode/electrolyte is more important. For mixed oxides, the diameter of the semicircle changes with the material composition [67], indicating that addition of a second metal oxide may improve the OER reaction kinetics. Moreover, in the case of $Ru_xIr_{(1-x)}O_2$ mixed oxides synthesized by a coprecipitation method in ethanol, synergetic effects were observed through lower R_{ct} values than those obtained on pure oxides [74]. Tafel slopes can also be drawn from impedance spectroscopy measurements [108,109]. The charge-transfer resistance is proportional to the current, so the anodic Tafel slopes can be extrapolated from the E vs log R_{ct}^{-1} curves knowing the R_{ct} values for the different potentials. The Tafel slopes are then determined from the linear parts of the curves (Fig. 12H). Conversely to the determination method of Tafel slope from polarization curves, that based on EIS measurements is not affected by the ohmic drop because only the charge transfers are considered. The Tafel slope values determined from EIS method are somehow very close to those determined from polarization curves, i.e., close to 40 mV dec^{-1} [74].

The determination of the stability of materials towards the OER and the evaluation of their long-term performances are generally based on repetitive CV measurements [52,53,67,107]. The first method consists in recording and comparing the shape of the CVs over the pseudo-capacitive potential range and to evaluate the charge (q^*) values as a function of the number of voltammetric cycles [52,53]. Wu et al. [67] have used this method to show that addition of SnO_2 to RuO_2 should increase the stability of the electrodes. In other works, the stability is determined from chrono-potentiometric measurements by applying different OER current densities after cycling. The rate of the increase of the potential vs time is an indicator of the stability of the catalytic material [79]. The less the potential value increases with time, the more stable is the electrode material [79].

3.3.1.2 Cathodic Electrode Materials
The researches in PEM electrolysis systems are mainly focused on the improvement of the anodic materials because the kinetics of HER at the cathode is much faster than those of OER at the anode, and the HER occurs with low overpotentials on platinum and palladium.

Carbon-supported Pt nanoparticles are actually the benchmark electro-catalysts for cathode materials in PEMEC [90]. Recent works have been devoted to the development of innovative material for HER, but the activity and stability during long-term tests were not efficient enough, so more researches are needed before replacing platinum for the HER in PEMEC.

Although few specific studies were devoted to Pt- and Pd-based cat-alysts for the HER in acidic media, these catalysts are the same as those used for the HER in alkaline media and for the oxygen reduction reaction in PEM fuel cells.

3.3.1.2.1 Physicochemical Analysis

A wide range of methods have been developed for the synthesis of noble metal based catalysts dispersed on a carbon conductive support: physical methods such as plasma sputtering of metals [110], laser abla-tion [111], MOCVD [112], etc., electrochemical methods [113], chemi-cal methods such as the impregnation-reduction method [114], colloidal methods [115], etc. The goal is to obtain nanoparticles with size in the range of 2 to 5 nm, well dispersed on the carbon support in order to achieve the higher number of active sites as possible.

One of the most important characterizations consists in determining the actual metal loading on the carbon support. Thermogravimetric analyses are very convenient for this purpose. Fig. 13A shows the ther-mogram and the weight derivative curves obtained under air flow for a Pt/C catalyst synthesized by a derived Bönneman method [116] with a nominal Pt ratio of 40 wt%. A first decrease of weight (lower than 5%) between 300 K and 400 K is attributed to desorption of physically adsorbed impurities and/or water from the surface of the carbon sup-ported catalyst. From 400K to 500K, the second small weight loss is attributed to desorption of chemically bounded water molecules. Then, in the range from 600 K to 800K, the combustion of carbon support takes place. At last, for temperatures higher than 800 K, the sample weight remains constant and the final weight percentage is attributed to the remaining metallic particles. The total weight loss is approxi-mately 60%, which corresponds to a Pt loading of 40 wt%. Lankiang et al. [117] used this method to determine the loading of Pd- and

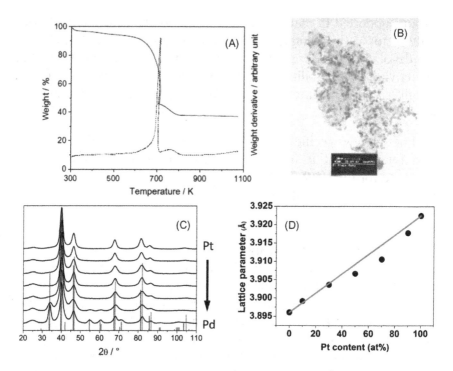

Figure 13 (A) TGA and weight derivative curves recorded under airflow at 5 K min^{-1} and (B) TEM images for a Pt(410 wt%)/C catalyst prepared by a derived Bönneman method [116]. (C) XRD patterns and (D) graph of the lattice parameters as a function of the catalyst atomic ratios for Pt$_x$Pd$_{(1-x)}$/C catalysts prepared by a "Water in Oil" method [117].

Pt-based catalysts prepared by a colloidal method and obtained actual loadings in very good agreement with the nominal ones.

The morphology of the supported catalysts, the particle size, and the size distribution are determined using TEM measurements. Fig. 13B presents a typical TEM image of Pt/C catalysts synthesized by a derived Bönneman method. Nanoparticles present a well-defined uniform round shape and are well dispersed on the carbon substrate, with a mean particle size of ca. 3.7 nm. Grigoriev et al. [118] obtained platinum clusters with a diameter between 2.5 and 3.5 nm uniformly distributed on a carbon support. For the HER, the requirement for the catalyst is to have the higher number of surface active sites where the reaction takes place. The electroactive surface area (EASA) of the catalyst is not only related to the particle size, but also to the dispersion of the nanoparticles on the surface. The development of synthesis methods allowing high dispersion of nanoparticles is very important, as well as the development of high specific area electron conductive support; the higher the support specific surface area, the higher the catalyst dispersion. The polyol method assisted by microwave irradiation is a very convenient method to obtain Pt nanoparticles of ca. 3 nm diameter well dispersed on a high surface area carbon black (Vulcan XC 72, c. 250 m^2 g^{-1}) leading to a value of the active surface area of 80 m^2 g^{-1} [119]. The value of the active surface area allows calculating the electrochemical particle size (d_{Elec}) according to the following equation [116]:

$$d_{Elec} = \frac{6000}{\rho_{Pt} * EASA} \tag{64}$$

where EASA is the electroactive surface area (80 m^2 g^{-1}), and ρ_{Pt} is the platinum density (21,800 kg m^{-3}). A value of 3.44 nm, close to that of 3.0 nm determined from TEM measurement, is obtained, which indicates a low agglomeration level of the Pt particles.

XRD measurements are used to determine the crystallographic structure of the particles, the mean crystallite size, the presence of alloys, and their compositions. Lankiang et al. [117] showed that the diffraction patterns of Pt$_x$Pd$_{(1-x)}$/C materials synthesized by a "water in oil" method always displayed the characteristic peaks of the face-centered cubic (fcc) structure of metallic particles. However, a shift of the peaks is observed as a function of the composition owing to the difference in the lattice parameters of pure platinum and palladium (Fig. 13C). The lattice parameters of the Pt$_x$Pd$_{(1-x)}$ materials can be calculated as a function of the composition using the Bragg equation and the plot of the lattice parameter *vs* Pd at% can be drawn (Fig. 13D). A straight line was obtained indicating that the Vegard's law was respected, and further that alloys were formed between both Pt and Pd metals. The mean crystallite sizes were calculated using the Scherrer equation and compared to the mean particle sizes from TEM measurements to verify if nanoparticles were mono- or polycrystalline. Mean crystallite sizes were of the same order as mean particle sizes for all Pt$_x$Pd$_{(1-x)}$ sample, indicating that nanoparticles were monocrystalline.

In the case of supported multimetallic nanoparticles, supplementary physicochemical analysis can be performed as for example XPS measurements, atomic absorption spectroscopy analysis, electrochemical characterizations, etc. in order to determine the surface states of the catalysts and interaction between the different metals [117].

3.3.1.2.2 Electrochemical Behavior

Typical CVs of Pt/C and PtPd/C electrocatalyst are presented in Fig. 14A. The potential region between 0.05 and 0.4 V *vs* RHE corresponds to the adsorption (cathodic scan) and desorption (anodic scan) of the atomic hydrogen on the metallic surface. The potential region between 0.4 and 0.6 V *vs* RHE corresponds to the capacitance double layer region. At higher potentials, oxidation of the metal surfaces occurs for the anodic scan, whereas the reduction of the formed oxides takes place for the cathodic scan [117,120]. Same results were

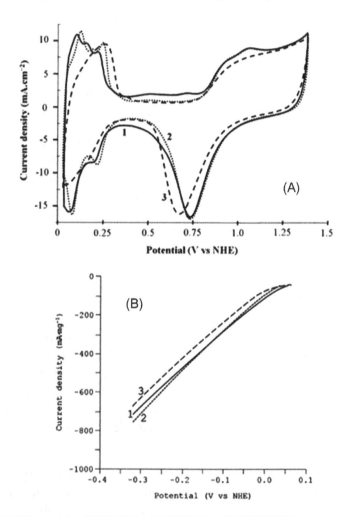

Figure 14 (A) CVs recorded on Pt40/XC-72 (1), Pt40/GNF (2), and PtPd40/GNF (3) electrodes in 1M H_2SO_4 at 25°C at a scan rate of 20 mV s^{-1} for a catalyst loading of 1 mg cm^{-2}; (B) related polarization curves measured on catalysts deposited on vitreous carbon electrodes at 80°C and a scan rate 1 mV s^{-1} [118].

obtained by Grigoriev et al. [118,121] with materials having similar particles sizes.

The EASA of catalysts containing platinum or palladium are obtained by integration of the CV curves in the potential of H desorption to determine the charge associated to the desorption of hydrogen (Q_{Hdes}) (integration of CVs curves between 0.05 and 0.4 V vs RHE in the case of Pt/C). Then, the capacitive contribution is subtracted in order to keep only the faradaic contribution. In the case of a Pt/C catalyst, the EASA values are determined using the following equation [121]:

$$EASA\left(m^2g^{-1}\right) = \frac{Q_{Hdes}}{Q_{H_0}m_{Pt}} \tag{65}$$

where Q_{H_0} is the charge for the desorption of a hydrogen monolayer from a smooth and flat platinum surface ($Q_{H_0} = 210\ \mu C\ cm^{-2}$) [122] and m_{Pt} is the platinum mass in electrode. The electrochemical performances of a catalyst for the HER are directly linked with the EASA.

The electrocatalytic activity of the catalysts towards the HER is evaluated from polarization curves [118,121]. Typical polarization curves are given in Fig. 14B for a Pt/C and PtPd/C electrodes. Generally, the current densities are normalized with respect to the noble metal mass in the catalytic layer. As for OER measurements, the onset potential of hydrogen evolution and the overpotential at a specific current density can be determined from the polarization curves, allowing one to compare the performance of different catalysts. Tafel plots can also be drawn from polarization curves in order to determine the rds of the reaction and the exchange current density. The most common HER mechanism in acidic media involves two steps—the first one being the hydrogen adsorption on the active site of the electrode (Volmer reaction):

$$H^+ + e^- + * \rightarrow H^* \tag{66}$$

where * is the active site, H^* is the adsorbed hydrogen atom on the active site. For this reaction, the proton comes from the electrolyte and the electron from the electrode. Then, two different pathways were proposed for the HER [123–125]. The first one is the Heyrovsky reaction:

$$H^* + H^+ + e^- \rightarrow H_2 \tag{67}$$

In this case, the adsorbed hydrogen intermediate reacts with one electron and one proton from the electrolyte to form one H_2 molecule. The second pathway corresponds to the Tafel reaction where two adjacent adsorbed hydrogen intermediates react to form one H_2 molecule:

$$2H^* \rightarrow H_2 \tag{68}$$

The Tafel slope associated with each step being the rds should be 120, 40, and 30 mV dec^{-1} for reactions (66)–(68), respectively [125]. The lower is the Tafel slope, the higher is the activity towards the HER. In the case of spherical Pd nanoparticles, Zalineeva et al. [126] obtained two Tafel slopes, the first one at low overpotentials was 46 mV dec^{-1}, and the second one for high overpotentials was 121 mV dec^{-1}. On the basis of the Tafel plot analysis, they proposed that in the low overpotential region, HER follows the Volmer – Heyrovsky mechanism with the reaction (67) being the rate-determining step, whereas in the high overpotential region, the reaction (66) becomes the rds, but both the Volmer – Heyrovsky and the Volmer–Tafel mechanisms could be operational.

3.3.2 Electrolyte/PEM

The proton-conducting membrane plays an important role in the production of pure gases and in the durability of the system performances. Currently, PFSA polymer membranes, such as commercial Nafion membranes, are the most used solid electrolytes in PEMEC [52,65,79,118,121], owing to their excellent chemical and thermal stabilities, mechanical strength, and high proton conductivity [127,128]. The main drawbacks of these membranes are their high costs (ca. 400 $ per square meter), the presence of fluorine in the polymer structure, their thickness (which increases the ohmic resistance and decrease the cell performance at high current densities), and their mechanical strength loss at high temperature [129–131].

Such kinds of proton-conducting membranes are also used for PEM fuel cells; however, although the structure and the chemistry of the membranes are quite similar for both applications, the requirements and the environment during operation are different, particularly in the point of view of the their hydration [132]. In an electrolysis cell, the membrane is fully hydrated whereas in a fuel cell the membrane it can be partially hydrated according to the working point, this parameter can affect the membrane properties and performances [10]. Some works are made to improve the membranes properties or to find alternative materials. Nafion membranes have been modified with oxides particles to improve their thermal properties [133,134]. Alternatives hydrocarbon membranes, such as polybenzimidazoles, poly(ether ether ketones) (PEEK), poly (ether sulfones) (PES), and sulfonated polyphenyl quinoxaline, have been developed in order to decrease the cost of the materials [10].

For all these membranes, the proton-conductivity properties are provided by the presence of sulfonic acid groups. Fig. 15A presents the schema of the protons migration through the membrane. The repartition of sulfonic groups and the structure of the ionic clusters inside the membrane are very important parameters. Albert et al. [135] evaluated the distribution of the sulfonic groups in innovative membranes by scanning electron microscopy (SEM) analysis coupled with EDS (Fig. 15B). The sulfur concentration is more important on the membrane surface, so the grafting method does not allow a well-homogeneous repartition of the proton conductive groups. They also carried out strain−stress measurements to determine the mechanical properties of the membranes and compare with commercial Nafion membranes (Fig. 15D): Both membranes led to very close results.

The gas permeability properties of the membranes are also an important parameters for enhancing long-term stability of the electrolysis cell and for obtaining hydrogen and oxygen with very high purity. The permeability is generally determined by measuring the crossover current, although this method is difficult to realize in PEM water electrolysis cell owing to the high pseudo capacitive properties of anode materials. Albert et al. [135] performed the crossover study in PEM fuel-cell mode with fully hydrated gases (Fig. 15C). They used one electrode as counter and reference electrodes, and CV measurements were recorded on the other electrode at a low scan rate in order to minimize the capacitive contributions of the carbon support. The current density obtained at 0.4 V *vs* RHE was principally due to the oxidation of the hydrogen that crosses the membrane. The measurements were realized for different membranes, and the current densities at 0.4 V could be compared.

3.3.3 Cell Performances

Electrolysis cell performances are determined on a membrane electrode assembly (MEA) composed of a proton exchange membrane (PEM),

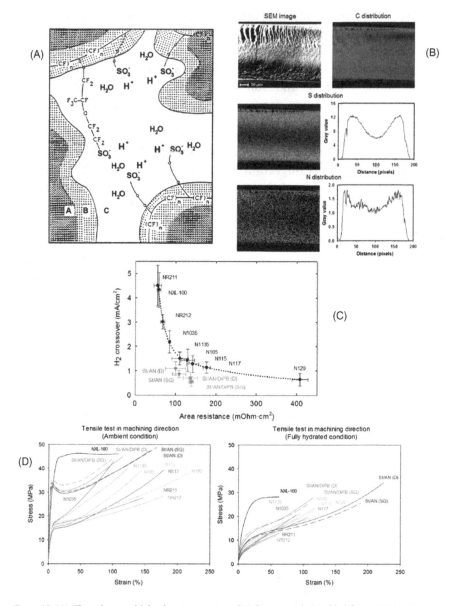

Figure 15 (A) Three-phase model for the microstructure of Nafion materials (modified from [118]); (B) SEM image and EDX mappings of a cross-section of St/AN/DiPB (D) membrane with a graft level of 42%; (C) property map showing hydrogen crossover vs area resistance values of various membranes, and (D) results of tensile tests of membranes in the machining direction under (left) ambient and (right) fully hydrated conditions [135].

an anode, and a cathode. For the electrode preparation, catalysts are dispersed in a water/alcohol solution in presence of a polymer binder (generally Nafion) to obtain a catalytic ink. The catalytic inks can either be deposited onto a gas diffusion layer or a current collector (catalyst coated backing = CCB) by spray, evaporation casting, etc., or on each side of the membrane (catalyst coated membrane = CCM) [136]. The interfaces between all MEAs component can eventually be improved by hot pressing. Rozain et al. [137] characterized IrO_2 anodic materials as a function of the catalyst loading by CV measurements (Fig. 16A). The Pt/C cathode in presence of hydrogen was used as counter and reference electrodes, whereas the IrO_2 anode was fed with nitrogen. The total charges, the most accessible charges, and the active site accessibility could then be determined in real electrolysis configuration using the equations of Ardizonne et al. [98]. The charges increase with the metal oxide loading and tend to stabilize at high loadings. Such measurements performed before and after operation of the MEA can be useful to evaluate the effect of operating conditions on the electrode materials.

An activation procedure can be applied before recording the performances of MEAs [137,138], but the main parameters, such as time, temperature, water flow, and applied currents, differ according to the research groups. The performances are generally evaluated by recording the stationary polarization curves (Fig. 16B) from 0 to ca. 2 A cm^{-2}. For each applied current density, the lower is the electrolysis cell voltage, the higher are the performance and the efficiency of the materials composing the MEA. Carmo et al. [139] listed the highest performances obtained in PEMEC: at 1 A cm^{-2} the cell voltage lies between 1.6 and 1.8 V, and at 2 A cm^{-2} between 1.7 and 2 V.

The electrochemical characteristics of the materials composing the MEA can also be determined by EIS. Nyquist diagrams can be drawn at different current densities or at the same current density for different MEAs in order to compare the effect of the modification of one element of the MEA (Fig. 16C). The shapes of the Nyquist diagrams can differ due to the acquisition mode of the experimental data or to the conception of the MEA. Rakousky et al. [143] observed only one semicircle, Mayousse et al. [140] obtained two semicircles, and Rozain et al. [142] presented intermediate results with two overlapped semicircles. In all cases, the semicircles were depressed certainly due to the roughness of the electrocatalytic layer and CPE were used to adjust the experimental data. When only one contribution is observed, the fitting procedure is applied with EEC composed of only one capacitive loop $R_\Omega \left(R_{ct} Q_{CPE(dl)} \right)$ where R_Ω is the cell resistance, R_{ct} is the charge transfer and $Q_{CPE(dl)}$ is the capacitance. From these results, only the

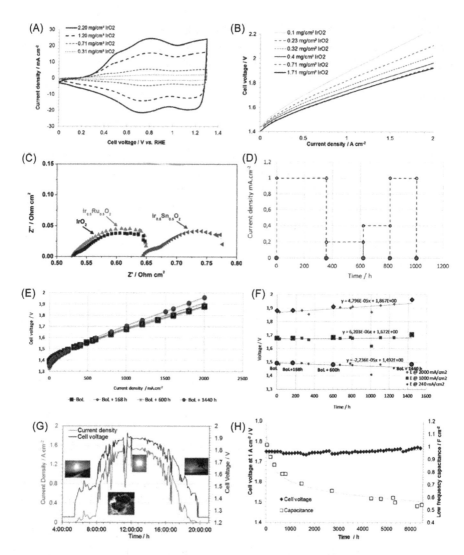

Figure 16 (A) CVs of anodes with various IrO_2 loadings at a sweep rate of $20\ mV\ s^{-1}$ and room temperature; (B) polarization curves at $80°C$ using different anode catalyst loadings and $0.25\ mg\ cm^{-2}$ Pt/C, and Nafion115 electrolyte membrane [137]; (C) impedance spectra acquired on the IrO_2, $Ir_{0.5}Ru_{0.5}O_2$, and $Ir_{0.5}Sn_{0.5}O_2$ oxides in the PEM water electrolysis cell at $j = 0.40\ A\ cm^{-2}$, atmospheric pressure and room temperature [140]; (D) input current signal of the ageing test; (E) changes in the polarization curve at $80°C$; (F) voltage slope at 240, 1000, and 2000 mA cm^{-2} as a function of time during aging test [138]; (G) current density profiles used for aging tests which is a real "solar"-type profile composed of a stand by phase, numerous, and frequent variations of current (cell temperature = $60°C$); (H) measured cell voltage evolution during solar-type aging test at 1 A cm^{-2} and $60°C$ on a 1.6 mg cm^{-2} pure IrO_2 MEA. Secondary axis: evolution of the low-frequency capacitance with aging time [141].

global capacitance and the global charge transfer can be evaluated. Although the time constants of the electron transfer are not the same for the OER and the HER, their separation is not possible in this configuration. Contrariwise, in the cases where two semicircle were observed, the contributions of both electrodes can be dissociated. The first semicircle at high frequencies can be attributed to the cathodic material and the second at low frequencies to the anodic material. So, the adjustments of the Nyquist plots were realized with two capacitive loops (one for each electrode contribution) and the EEC was $R_\Omega \left(R_{ct}^a Q_{CPE(dl)}^a \right) \left(R_{ct}^c Q_{CPE(dl)}^c \right)$. Mayousse et al. [140] showed that the cell resistance and the charge transfer were dependent on the metal oxide composition. Globally, the cell resistances measured by EIS are included between 100 mΩ cm^{-2} and 250 mΩ cm^{-2}.

Long-term measurements were also performed either under stationary or dynamic conditions. Ma et al. [64] performed galvanostatic measurements at 1.1 A cm^{-2} for 2000 h on MEAs with RuO$_2$ treated at different temperatures as anodic materials. The cell voltage decreased during the first step of the measurements, which was attributed to the wetting of the MEA or to the dissolution of the RuO$_2$ catalyst, then the cell voltage tended to stabilize. The highest performance was obtained for the material treated at 350°C. Fouda-Onana et al. [138] applied the aging test procedure presented in Fig. 16D. At the end of each sequence, polarization curves were recorded and compared (Fig. 16E). The cell-voltage degradations were estimated to be -22, $+6$, and $+47$ μV h^{-1} at 0.24 A cm^{-2}, 1 A cm^{-2} and 2 A cm^{-2}, respectively (Fig. 16F). The negative value for the potential degradation at 0.24 V was explained by the membrane thinning as confirmed by fluoride titration in exhaust water on the anode and cathode side. Rozain et al. [141] and Audichon et al. [74,93] performed long-term tests under galvanodynamic conditions and simulated the coupling electrolysis cell/solar cell. The solar power profile corresponding to a 14 hours sunny-day (Fig. 16G) was applied several times to the PEMEC. Many successive solar cycles were applied to evaluate the catalyst properties and MEA performances. A polarization curve was recorded after each cycle, and the value of the cell voltage at 1 A cm^{-2} was plotted *vs* time (Fig. 16H). Rozain et al. [141] demonstrated that the increase of the catalyst loading led to minimize the cell voltage degradation rate; a loading of 1.6 mg cm^{-2} allowed reaching a stable cell voltage for 6000 h experiment. Audichon et al. [93] showed that the addition of iridium to the RuO$_2$ OER catalyst led to decrease the cell voltage degradation. Rakousky et al. [143] recorded the changes in anodic and cathodic voltages for long-term measurements by integrating a dynamic hydrogen electrode reference. They pointed out that the cathodic overpotentitals were lower than the anodic ones. TEM analyses of the electrode materials before and after long-term tests showed particle growth and agglomeration for both the OER and HER electrocatalysts.

3.4 ALKALINE WATER ELECTROLYSIS

Currently, the alkaline technology is almost the only electrolysis process for the production of hydrogen used by industry. The size of electrolysis modules can be modulated as a function of the hydrogen demand rate, from ca. 0.5 to ca. 800 N m^3 h^{-1} [144]. The following sections will describe the different components of AECs.

3.4.1 Electrolyte/Diaphragm

The electrolyte of alkaline water electrolysis systems is an aqueous solution of potassium or sodium hydroxide. The potassium or sodium hydroxide concentrations, which can vary as a function of the working temperature, is generally in the 25 wt% to 30 wt% range for temperatures between 70°C and 100°C and pressures between 1 bar and 30 bars [145]. During cell operation, hydroxide ions are formed at the cathode by the reduction of liquid water into gaseous hydrogen $(2H_2O_{(l)} + 2e^- = H_{2(g)} + 2OH^-_{(aq)})$, hydroxyl ions migrate through the electrolyte towards the anode where they are oxidized into oxygen and water $(2OH^-_{(aq)} = 1/2O_{2(g)} + H_2O_{(l)} + 2e^-)$, water can retrodiffuse towards the cathode to be reduced. Because of the diffusion and migration processes of these species, and in order to separate the different gases formed at the cathode and anode, respectively, the use of a diaphragm is generalized in such alkaline systems, mainly for safety reasons.

A diaphragm is a microporous material with average pore sizes less than 1 μm, allowing the transport of water and hydroxyl ions between the anode and the cathode compartments, and the separation of gases. Fig. 17 shows the basic scheme of an alkaline water electrolysis cell. The characteristics required for the diaphragm are high permeation to water, high corrosion resistance in strongly alkaline media, and high ionic conductivity to obtain high cell efficiency [146]. Indeed, the AEC overvoltage during operation is not only due the overpotentials from HER and OER, but also to the ohmic loss in the electrolyte, which includes the resistance from ionic transfer in the diaphragm. This ohmic loss induced by the presence of the diaphragm explains the

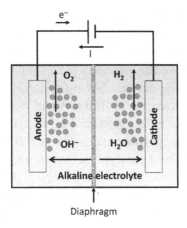

Figure 17 Basic scheme of an alkaline water electrolysis cell.

reason why lower current densities can be applied to AECs in comparison with PEMECs.

Several kinds of diagrams have been studied such as composite materials based on ceramic materials or microporous materials: reinforced microporous PES membranes, glass reinforced polyphenylene sulfide compounds, nickel oxide layers on a mesh with titanium oxide and potassium titanate [147]. The diaphragm should display high performance, low cost, and nonhazardous characteristics. NiO has shown to be a good material and some fabrication methods have been developed for controlling its thickness or porous structure [148].

3.4.2 Electrode Materials

As in the case of PEMEC, the anodic overpotential of the OER is also limiting for the development of efficient alkaline electrolysis systems. The best catalysts for this reaction are based on IrO_2 and RuO_2 materials, but the low stability of these oxides in alkaline media makes them unusable in such media [149]. Indeed, owing to the use of an aqueous alkaline electrolyte, the electrode materials have to be corrosion-resistant whereas keeping high catalytic properties *vs* time. The most used materials for AEC anodes are based on nickel, cobalt, and iron [145]. However, electrodes based on nickel, and particularly nickel electrodes recovered by a nickel oxide layer, show very good stability in alkaline media and are currently used in commercial systems [150]. Cobalt oxides of spinel structure are also studied for the OER [151], as well as mixed oxide of spinel structure (nickelite, cobaltite, and ferrite), the nanostructured cobalt cobaltite presenting the best catalytic performances [152]. On the other hand, the conductivity of Co_3O_4 species is generally low, and this material can be doped by Li and La to improve this property [153,154]. Concerning the mixed oxides, a nickel cobaltite aerogel $NiCo_2O_4$ catalyst has led to the very interesting OER performance of $100\ mA\ cm^{-2}$ at an overpotential of $184\ mV$ [155]. Chanda et al. studied $Fe_xNi_{1-x}Co_2O_4$ and found the highest electrocatalytic performances for $x = 0.1$ [156]. In order to improve the durability and catalytic activity by stabilization of the surface using an electroconductive oxide, $Li_xNi_{2-x}O_2/Ni$ electrodes have also been developed, which displayed good catalytic activity and durability during potential cycling [157].

Platinum nanoparticles deposited on a carbon black support can obviously be used for the HER at the cathode of an AEC. But in such media, cheaper materials also display relatively good activity for this reaction. Commercial alkaline electrolysis systems generally use mild steel recovered by a nickel layer as cathode. Andrade et al. have tested sixteen commercial alloys in terms of activity and stability and compared the results with those obtained on the most employed cathode material in diaphragm AECs, i.e., mild steel SAE1020. They found that 2RK65 (Ni 25.7 wt%, Mo 4.43 wt%, Cr 17.9 wt%, Cu 1.23 wt%), SAF2304 (Ni 4.5 wt%, Cr 23 wt%, Fe 69 wt%, Mn 2 wt%, Si 1 wt%), SAF2507 (Ni 7 wt%, Mo 4 wt%, Cr 25 wt%, Fe 61 wt%, Cu 0.5 wt%, Mn 1.2 wt%, Si 0.8 wt%), C22 (Ni 58.9 wt%, Mo 13.4 wt%, Cr 19.5 wt%, Cu 0.92 wt%, W 2.87 wt%), and MONELK500 (Ni 67.5 wt%, Cr 0.47 wt%, Cu 30 wt%, W 0.517 wt%, Ti 0.50 wt%) were advantageous to substitute the mild steel currently used [158]. The 316L stainless steel has also shown good properties for the HER [159], and it has been proposed that an electrolysis cell working with stainless steel anode, and cathode led to lower operational cost than classical commercial electrolysis systems although the electric energy consumption could be higher [160]. Here, the fluctuation of the cost of the electric energy provided to the cell will determine the kind of technology developed for such systems. Nickel-based electrodes have also been extensively studied [161]. But owing to the relatively high cost of nickel in comparison with steel, nickel catalysts have to be nanostructured or modified by alloying in order to increase the catalytic performance of the electrode towards the HER. For example, Baranton et al. studied the behavior of nanostructured Ni_xCo_{10-x}/C catalysts for the HER. They found that the catalytic activity increased regularly with the increase of the cobalt content in the Ni_xCo_{10-x}/C nanoflakes, but according to long-term stability measurements performed by chronoamperometry, dissolution of cobalt occurred. The best compromise between activity and stability was found for Ni_5Co_5/C and Ni_7Co_3/C catalysts [12].

3.4.3 Cell Performances

Recently, Schalenbach et al. [162] compared the efficiency of state-of-the art AEC and PEMEC. The AEC was fitted with a commercially available Zirfon separator with a thickness of approximately 460 μm in aqueous 30% KOH, with nickel-based electrodes. The electrolyte of the acidic water electrolysis cell consisted of a Nafion N117 membrane (a thickness of approximately 209 μm in the wet state) and Nafion ionomer-based electrodes, with Pt/C at the cathode and IrO_2 at the anode. Fig. 18 shows the polarization curves recorded at 80°C.

It was proposed that the lower efficiency of the alkaline water electrolysis cell was mainly due to the higher cell resistance, which mainly can be attributed to the employed macroporous electrodes and the thicker separator in comparison to the membranes of the acidic water electrolyzers.

Carmo et al. depicted the state-of-the-art for the specifications of alkaline and PEM electrolyzers [139], confirming that the current densities at given voltages were three to five times higher for a PEMEC than for an AEC.

3.5 HIGH TEMPERATURE WATER ELECTROLYSIS

More recently, high-temperature water electrolysis technology, working between 500°C and 1000°C, has known a growing interest in the scientific community. Indeed, the increase of the electrolysis cell working temperature allows decreasing the electric energy demand for running the system thanks to the high heat provided to the cell [163]. The electrochemical reactions taking place at such high temperatures are completely different from those occurring in PEMECs and AECs working at temperatures lower than 100°C. In high-temperature electrolysis cells the cathode is fed with steam water, which is reduced into hydrogen and oxide ions (O_2^-). The oxide ions migrate towards an anion conducting electrolyte and are oxidized at the anode to produce oxygen [164].

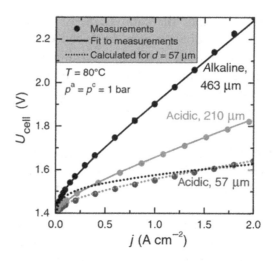

Figure 18 Polarization curves of the alkaline cell (black) and the acidic cells with a Nafion N117 membrane (wet thickness of 210 μm, graphed in red (in online version)) and a Nafion NR212 membrane (wet thickness of 57 μm, graphed in blue (in online version)). The black and red dotted lines show modeled voltage-current characteristic used with the same parameters as the solid black and red lines, respectively, but assuming separator thicknesses of 57 μm, Re = 0 (Efficiency analysis section) for the alkaline cell and Y = 2 bar cm² A⁻¹ for the acidic cell (gas crossover section) [162].

Owing to the high working temperature of the electrolysis cells, the components have to be thermally stable and are then made of solid oxides. Such systems are then called solid oxide electrolysis cells (SOECs). Electrodes and electrolyte are then very similar as those used in solid oxide fuel cells (SOFC). This electrolysis technology has then benefited from the numerous researches performed on the development of materials for SOFC.

The electrode materials have to be electron conducting porous ceramic in order to facilitate electron and mass transport (oxygen and hydrogen), as well as ionic conducting in order to allow migration of O_2^- species. Electrodes materials consists generally in conducting mixed oxides with perovskite structure (ABO_3). The mix of different metals, with different doping levels, has been performed with the goal of improving both activity and stability [165,166]. The composition of the anode catalysts is mostly based on lanthanum, strontium, and manganese whereas the composition of cathode catalysts is mostly based on nickel and zirconium [167]. The solid electrolyte is generally a mixed oxide based on yttrium oxide stabilized by zirconium oxide (ZrO_2/Y_2O_3) [163].

The main degradation processes are induced by the working conditions of the system, and particularly the high temperature and the presence of the water steam. Moreover, the different components of the membrane-electrodes assembly display different thermal dilatation coefficients; therefore, start-up/shut-down cycles are detrimental for the interfacial contact between electrodes and electrolyte, leading to delamination of the different layers and increase of the cell internal resistance, and further to a loss of performances [167]. In addition, in order to increase the conduction properties, the different components have to be very thin, which makes them very fragile, and owing to their nature the MEA is very rigid, which makes the assembling of the cell very complex.

For these reasons, before this technology could reach a commercial deployment, numerous researches have still to be devoted to the optimizations of materials, components, and systems. But such technology has a real potency as hydrogen production system by dissipating the excess of heat (and reaching the working temperature of ca. 800°C) and the electricity produced by concentration solar power plants or nuclear power plants. It could then allow storing chemical energy for restituting electric energy to the grid through a fuel cell, participating then to the development of the energetic mix.

3.6 LIMITATIONS AND PERSPECTIVES OF WATER ELECTROLYSIS

Although AECs are the most developed systems for practical applications, the development of very active dynamic stable electrodes based on noble metals (ruthenium and iridium) and of highly proton-conductive PFSA membranes made acidic electrolysis much more efficient than alkaline systems. PEMEC systems also offer high compactness and a very short start-up time even at room temperature.

In PEM water electrolysis cells, water molecules are oxidized at the anode generating oxygen and protons. Protons migrate through the polymer electrolyte membrane and at the cathode they are reduced into hydrogen. The proton-exchange membrane electrolyte has to be hydrated to allow efficient proton conduction; it also plays the role of gas separator allowing the production of pure hydrogen. This technology can operate in the temperature range where water is under liquid phase (lower than 100°C). The main limitations of the proton exchange membrane water electrolysis cells are the high cost of the different elements and their stability for long-term hydrogen production. The research activities are then principally focused on the decrease of the production cost of electrolyte membranes, the diminution of the amount of strategic, and costly platinum group metals in electrodes, while maintaining or even improving the stability of the whole system.

AECs operate at low temperatures (typically below 80°C). The anode and the cathode electrodes are immersed in an alkaline media, the most commonly used being a potassium hydroxide solution. The water molecules are reduced at the cathode forming hydrogen and hydroxyl ions (OH⁻). Hydroxyl ions migrate through a diaphragm from the cathodic side to the anodic side and are oxidized at the anode forming oxygen and water. The diaphragm has the property to be permeable to OH⁻ species and also acts as a separator for both gases produced. The development of thinner diaphragm is a way to increase the efficiency of this technology, particularly by decreasing the ohmic loss between the electrodes. Currently, the alkaline electrolysis is the most employed electrochemical process for hydrogen production in the industry.

SOECs operate generally over a temperature range from 500°C to 1000°C. So, water is under gas phase, and a water steam is injected in the cathodic side where it is reduced in hydrogen and O^{2-} oxide species. O^{2-} oxide species migrate through a solid electrolyte, generally zircone yttrium based solid electrolytes. At the anode side, O_2^- species are oxidized in oxygen. Due to the high operating temperature of such systems, all components of the electrolysis cells (electrodes, electrolyte, current collectors, etc.) are solid and should be thermally stable.

Regarding the operation conditions, or of the reactions evolving at the electrode surface, the materials used in these different systems need to have specifics properties. For all these reasons, each types of electrolysis cells can present some advantages and drawbacks, as summarized in Table 4.

The main advantages for alkaline electrolysis systems are the possibility to use non-noble electrocatalyst materials and its long-term stability in a stationary mode. But the highly corrosive electrolytic media can lead to corrosion of the external system. Moreover, the diaphragm using as separator cannot avoid totally the gas permeation during long-term hydrogen production. At last, the lower mobility of hydroxyl ions compared with protons limits the conductivity and further the cell efficiency.

Table 4 Main Advantages and Drawbacks of Electrolysis Cell Systems [163,168]			
	Alkaline Electrolysis	**PEM Electrolysis**	**Solid Oxide Electrolysis**
Advantages	Commercial technology	Near-term technology	Efficiency up to 100%
	Non-noble catalysts	High current densities	Thermoneutral voltage
	Long-term stability	High voltage efficiency	Non-noble catalysts
	Relatively low cost	Good partial load range	High-pressure operation
	MW range stacks	Rapid system response	
	Cost effective	Compact system design High gas purity Dynamic operation	
Drawbacks	Low current densities		
	Crossover of gases	High cost of components	Mediate-term technology
	(degree of purity)	Acidic corrosive medium	Bulky system design
	Low partial load range	Possibly low durability	Durability (brittle ceramics) no dependable cost information
	Low dynamics	Stacks below MW range	
	Low operational pressures		
	Corrosive liquid electrolyte		

In PEMECs, the compactness of the system allows operating at high current densities while preserving the high purity of the produced hydrogen. Their rapid voltage response induced by the nature of the electrolyte (solid) and the species migrating through the electrolyte (protons) allows coupling these systems with intermittent renewable energy sources. However, the main drawbacks for this type of electrolysis systems remain their high cost induced by the components, such as platinum group metals in the electrodes, perfluorosulfonic membrane as electrolyte and porous titanium as current collectors, and their relatively low stability in the acidic environment.

Both these systems, AEC and PEMEC, can also be adapted for high-pressure electrolysis, allowing compressing hydrogen around 120–200 bars and further to avoid the use of an external hydrogen compressor.

For solid oxide electrolysis cells, the high operation temperature leads to a decrease of the electric energy consumption for producing hydrogen and allows using non-noble metals in ceramic oxides serving as electrodes and electrolyte. Nevertheless, the compactness of the systems and the thinness of the ceramic materials required for developing these systems make the MEAs easily to damage. Moreover, cracking can occur during long-term hydrogen production. But, due to high operation temperature requirement, such technology is quite interesting for the development of reversible unitized fuel-cell technology.

In spite of all the different drawbacks related to each electrolysis system for hydrogen production, each technology can have an application according to the considered domain: PEMECs can specifically be coupled with renewable energy systems, such as wind, solar, tidal powers, etc., AECs can be used for industrial stationary applications and SOECs for the valorization of the energetic production excess (heat and electricity) of nuclear power plants.

CHAPTER 4

Hydrogen Production From Biomass Electroreforming

The electroreforming of biomass can be performed in aqueous media by oxidation of organic compounds with coproduction of hydrogen on the cathode of the electrolysis cell. An advantage of compounds originating from biomass is that most of them are present as oxygenated molecules (alcohols and sugars) and can then be activated through electrochemical processes. Considering simple alcohols, polyols, and sugars as examples, Eqs. (69)–(73) describe the global reactions obtained in an electrolysis cell performing the complete oxidation of these compounds to CO_2 and water reduction at the cathode.

Methanol

$$CH_3OH + H_2O \rightarrow CO_2 + 3\,H_2 \tag{69}$$

Ethanol

$$CH_3 - CH_2OH + 3\,H_2O \rightarrow 2\,CO_2 + 6\,H_2 \tag{70}$$

Ethylene glycol

$$CH_2OH - CH_2OH + 2\,H_2O \rightarrow 2\,CO_2 + 5\,H_2 \tag{71}$$

Glycerol

$$CH_2OH - CHOH - CH_2OH + 3\,H_2O \rightarrow 3\,CO_2 + 7\,H_2 \tag{72}$$

Glucose

$$C_6H_{12}O_6 + 6\,H_2O \rightarrow 6\,CO_2 + 12\,H_2 \tag{73}$$

The electroreforming of organic compounds presents the advantage that the oxidation occurs in a separate compartment to that of water reduction into hydrogen. It makes this process an attractive way of producing pure hydrogen without release of carbon monoxide even in the case of a partial oxidation in the anodic compartment. Furthermore, the electroreforming of biosourced molecules is

Hydrogen Electrochemical Production. DOI: http://dx.doi.org/10.1016/B978-0-12-811250-2.00004-2

performed in aqueous medium with water as reactant on the cathode and is then performed at lower temperature (typically below 100°C) than classical chemical routes (steam reforming or partial oxidation).

4.1 HYDROGEN REFORMING FROM ALCOHOLS/POLYOLS

Among the large amount of molecules available from biomass processing or as byproducts of biomass industry, alcohols are well represented molecules. Ethanol is obtained after fermentation of biomass, ethylene glycol is obtained by acidic hydrolysis, and glycerol is a byproduct of biofuel industry. Hence, the formation of hydrogen from biosourced alcohols would represent a renewable route for hydrogen production, and the use of small molecules containing two to three carbons allows an easier electrochemical activation and a better control of the reaction selectivity than the use of heavier molecules such as sugars. These molecules have been envisaged as an alternative to hydrogen in fuel cells with the introduction of direct alcohol fuel cells [20,169–172], but their oxidation then had to comply with the constrains of fuel-cell performance requirements (in terms of power density and energy density) and have shown lower performances than hydrogen fuel cells. These works have paved the way for the use of alcohols in electroreforming process that avoids the constraints of fuel cells and by demonstrating the thermodynamic and electrocatalytic advantages of an alcohol anode as counter electrode for water reduction.

4.1.1 Thermodynamic of Alcohols/Polyols

The Gibbs energies of alcohol electroreforming reaction (ΔG_r°) are very low compared to that of water electrolysis (Table 1). From the Gibbs energy of the reaction, the equivalent thermodynamic standard cell voltage (U_{cell}^0) can be defined as

$$U_{cell}^0 = \frac{\Delta G_r^{\circ}}{nF} \tag{74}$$

From the reaction enthalpy (ΔH_r^0), the thermoneutral cell voltage (U_{cell}^{th}) can be defined as

$$U_{cell}^{th} = \frac{\Delta H_r^0}{nF} \tag{75}$$

Both these thermodynamic voltages define the working condition for an alcohol electroreforming cell depending on the cell voltage (U_{cell}):

- if $U_{cell} < U_{cell}^0$, the electroreforming reaction cannot occur;
- if $U_{cell}^0 < U_{cell} < U_{cell}^{th}$, the electroreforming reaction occurs, and extra heat is required to perform the reaction;
- if $U_{cell}^{th} < U_{cell}$, the electroreforming reaction occurs and release heat.

The values of U_{cell}^0 are low (close to 0 V) for alcohol electroreforming (Table 1) that implies that the reaction can occur at low cell voltage, and the expected open circuit voltage of an electroreforming system is expected to be close to 0 V. More importantly, the values of the thermoneutral cell voltage (U_{cell}^{th}) lies in the range from 0.2 V to 0.3 V, which indicates that an alcohol electroreforming system can perform hydrogen production at cell voltages up to four times lower than that of a water electrolysis cell. These low cell voltages originate from the energy brought to the system by alcohol consumption.

The electrical energy consumption of an electroreforming cell (We) working at constant cell voltage can be expressed as:

$$We = U_{cell} \int i(t)\mathrm{d}t = U_{cell}\, Q \tag{76}$$

where Q is the charge involved during the electroreforming process.

Independently on the alcohol considered, the water-reduction reaction occurring at the cathode of the electroreforming cell is

$$2\,H^+ + 2e^- \rightarrow H_2 \text{ (acidic medium)} \tag{77}$$

$$2\,H_2O + 2e^- \rightarrow H_2 + 2\,OH^- \text{(alkaline medium)} \tag{78}$$

The charge involved during the electroreforming process can then be expressed as:

$$Q = n_{e^-} F = 2n_{H_2}F = 2\frac{V_{H_2}}{V_m}F \tag{79}$$

where n_{e^-} is the number of mole of electron exchanged per mole of hydrogen produced (n_{H_2}), F is the Faraday constant, and V_m is the hydrogen gas molar volume. Combining Eqs. (76) and (79) allows the determination of the electrical energy consumption per volume of hydrogen produced:

$$\frac{We}{V_{H_2}} = \frac{2U_{cell}F}{V_m} \tag{80}$$

Eq. (80) evidences that the parameter affecting the electrical energy consumption per volume of hydrogen produced is the electroreforming cell voltage, U_{cell}. Hence, a cell operating in the range from $U_{cell} = 0.5$ V to 1.0 V would lead to an electrical consumption in the range from $We = 1.2$ kW h N $m_{H_2}^{-3}$ to 2.4 kW h N $m_{H_2}^{-3}$. Compared to water electrolysis, working at cell voltages in the range from 1.65 V to 1.8 V, leading to an electrical energy consumption from 3.9 kW h N $m_{H_2}^{-3}$ to 4.3 kW h N $m_{H_2}^{-3}$, the alcohol electroreforming requires two to three times less electrical energy for hydrogen production. This energy saving is obtained at the cost of the consumption of alcohol, which has several implications. At first, the alcohol should be obtained from a renewable source, and its consumption should not increase the production price of hydrogen over that of hydrogen from water electrolysis. It was demonstrated from electroreforming systems operating at 80°C and in the case of ethanol that cost implied by the complete loss of ethanol was overcompensated by the decrease of electrical energy consumption [173]. Furthermore, from this observation, two strategies arose: The first one consisted in yielding the maximum amount of hydrogen per reactant consumed in the electroreforming process (i.e., reaching the highest oxidation state possible by forming carbon dioxide from alcohol) [174−176]; the second one consisted in using valuable organic compounds to achieve the coproduction of hydrogen and value-added chemicals from the oxidation of alcohol [177,178]. These two strategies are made possible by the wide potential range available at the anode for alcohol electroreforming allowing a control of the selectivity of the oxidation reaction of the alcohol through the control of electroreforming cell potential (the lower the potential, the lower the oxidation degree of reaction products) as well as the catalytic materials used (for a control of the selectivity towards a given reaction product). Further, alternatives to water electrolysis for electrochemical hydrogen synthesis (due to the low kinetics of water oxidation at the anode and hence high energy consumption), such as conventional alcohols, bio-sourced alcohols and other compounds from biomass are discussed.

4.1.2 Electrocatalytic Aspects

The cell voltage is the key parameter for the production of hydrogen at low-energy cost [Eq. (80)], and the use of an alcohol allows reducing the electroreforming cell voltage at least from a thermodynamic point of view. In order to maintain the cell voltage as close to the thermodynamic value as possible, it is important to activate the alcohol oxidation reaction with an electrocatalytic material at the anode. Indeed, compared to water oxidation reaction, the alcohol oxidation reaction involves more complex mechanisms, and then more reaction steps to reach the reaction products. Considering the half-reaction equations of ethanol oxidation for example, the three following equations can be established depending on the reaction product:

$$CH_3-CH_2OH \rightarrow CH_3-CHO + 2\,e^- + 2\,H^+ \qquad (81)$$

$$CH_3-CH_2OH + H_2O \rightarrow CH_3-COOH + 4\,e^- + 4\,H^+ \qquad (82)$$

$$CH_3-CH_2OH + 2\,H_2O \rightarrow 2\,CO_2 + 12\,e^- + 12\,H^+ \qquad (83)$$

Fig. 19 presents an example of reaction pathways possible for ethanol oxidation in acidic medium corresponding to Eqs. (81)–(83) leading to the three stable reaction products (namely acetaldehyde, acetic acid, and carbon dioxide) and one minority product (methane).

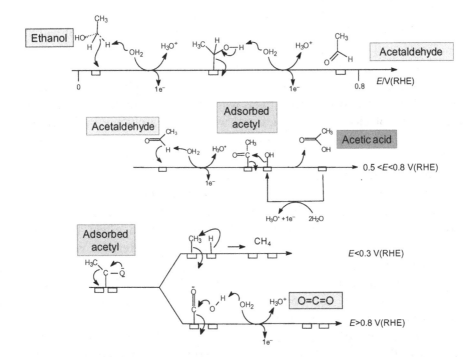

Figure 19 Possible reaction pathways for ethanol oxidation reaction in acidic medium as a function of electrode potential applied [179].

The half-reaction equations as well as the reaction mechanism evidence two important issues that have to be addressed from an electrocatalytic point of view in order to maintain an anode potential close to the thermodynamic potential: (1) The oxidation degree of reaction products and consequently the number of hydrogen molecules formed from ethanol electroreforming increases with the potential increase; (2) the coactivation of water together with ethanol is required to reach the highest oxidation degrees for reaction products.

The electrocatalysis of alcohol oxidation will then be focused on two main aspects: the catalyst activity in order to reach the highest reaction currents (i.e., the highest hydrogen production rate) at the lowest potential and the catalyst selectivity in order to obtain the desired oxidation degree of reaction products in the lowest potential domain possible. To meet these goals, several catalytic material designs can be employed. At first, the catalyst can be able to perform a bifunctional mechanism with the activation of both alcohol and water on its surface. The bifunctional mechanism is generally observed on metal alloys, PtRu being the most typical for alcohol oxidation [180], with platinum performing the adsorption and activation of alcohol molecules and ruthenium performing the adsorption and activation of water molecules at lower potential than platinum. The second effect affecting the catalyst performance is a geometric effect, obtained by alloying a catalytic material with another metal in order to change the lattice parameters. The geometric effect will then have an impact on the adsorption of alcohol molecules and will induce a change in the catalyst selectivity, and may also induce an increase of catalytic activity. Alloying the catalytic metal with a second metal may also induce changes in the electronic configuration of the alloyed material compared to the catalytic metal alone. This electronic effect can induce changes in electron transfer for the alcohol oxidation reaction as well as in adsorption of alcohol molecules and can then affect both the activity and the selectivity of the anode catalytic materials. While the bifunctional mechanism is generally easily identified and can be anticipated from the behavior of metals involved in the alloyed catalytic material when they are considered separately, it remains difficult to discriminate between a geometric effect and an electronic effect for the improvement in activity and selectivity observed from an alloyed material. An alloy such as PtSn for ethanol oxidation is a typical example of activity and selectivity changes in the catalytic activation of ethanol compared to Pt [18]. Compared to a Pt electrode, the use of PtSn as catalytic materials induces an orientation of the selectivity towards higher oxidation degree reaction products (i.e., the increase of the amount of carbon dioxide and the decrease of the amount of acetaldehyde in reaction products). This increase in the oxidation degree leads to a higher amount of exchanged electrons per ethanol molecule oxidized, also increasing the oxidation current measured. This change in activity is not attributed to a bifunctional mechanism, while not discriminating between an electronic effect, a geometric effect, or both being involved. The selectivity can also be controlled by adatoms [181]. The adsorption of a metal on the surface of catalytic materials will induce a modification of the apparent geometry of adsorption sites. This modification of the adsorption site may induce a modification of the reactant adsorption mode and consequently will impact the selectivity of the reaction.

The example in Fig. 19 presents an aspect of the reaction mechanism that can induce changes in the apparent selectivity with acetaldehyde: the possibility for a reaction product to be desorbed and released in the final reaction product or to be readsorbed and to react again to form reaction products having a higher oxidation degree. This part of the reaction mechanism can be controlled by controlling the adsorption strength of reaction intermediates (either through a modification of the catalytic material inducing a change in the electronic configuration of the material or in the geometry of surface sites) or by modifying the contact time of reactants with the catalyst. This latter parameter can be controlled independently on the operating parameters of the electroreforming cell (reactant flow rates, electrode surface, etc.) by changing the nanostructure of the catalytic material. For example, inducing a higher microporosity of the catalytic material may induce a longer residence time of the reaction products inside the porosity of the material and may then increase their chance to be oxidized again towards higher oxidation degree products. The nanostructure of the catalytic material can then play an important role for the selectivity and activity of the anode.

4.1.3 Electrode Materials
The nature of electrode materials used in electroreforming cells will depend on several parameters: the nature of the electrolyte (acidic or alkaline), the alcohol considered for the oxidation reaction at the

anode, and the oxidation reaction products targeted. For the cathode, the materials used are the same as that used in the case of acidic or alkaline water electrolysis. Considering the working potential range of the anode (below 1.0 V) as well as the temperature range (below 100°C), the durability of materials is not a parameter of major concerns. Hence, electrocatalysts are generally dispersed on a carbon support providing sufficient chemical and thermal stabilities. The nature of the anodic electrocatalysts is however governed by two other parameters: their activity for the alcohol oxidation reaction in order to reach the lowest anode potential possible to minimize the electrical energy consumption; their selectivity towards the formation of desired reaction products, either to reach a high oxidation with the formation of carbon dioxide or to maximize the yield of production of partially oxidized value-added compounds.

4.1.3.1 Methanol and Ethanol

Methanol and ethanol are the two simplest alcohols bearing a single primary alcohol function with one or two carbon. The electroreforming of these two alcohols is generally performed in a proton exchange membrane electrolysis cell (PEMEC) [174–176,182,183] and the strategy consists in recovering the maximum hydrogen molecules from their reforming through a complete oxidation of these alcohols into CO_2. Indeed, in the case of these alcohols, the intermediate oxidation products (i.e., aldehyde and carboxylic acid) are not considered as high value-added compounds, and it is then more interesting to produce a maximum of hydrogen molecules from these alcohols than to obtain less hydrogen with coproducts. Moreover, in the case of ethanol, acetaldehyde is a toxic compounds. The catalytic materials envisaged at the anode of these systems are developed with the idea of maximizing the activity and the selectivity towards carbon dioxide formation in acidic medium. In an acidic medium, platinum-based catalysts have demonstrated the highest activity for alcohol oxidation through the development of direct alcohol fuel cells [184,185]. However, for the formation of carbon dioxide from these two alcohols, carbon monoxide will be involved as reaction intermediate, and this compound is known to adsorb strongly on platinum surface and can only be desorbed through its oxidation into CO_2 [Eq. (84)].

$$Pt-CO + H_2O \rightarrow Pt + CO_2 + 2e^- + 2H^+ \qquad (84)$$

This reaction requires water activation, which is possible at potential over 0.8 V *vs* reversible hydrogen electrode (RHE) on platinum, leading to a high working cell voltage. The use of a second metal such as ruthenium allows decreasing the cell voltage by decreasing the water activation potential on ruthenium surface. In order to benefit from both the catalytic activity of platinum for alcohol oxidation and the catalytic activity of ruthenium for water activation (bifunctional mechanism), PtRu alloys are generally used in anodes for methanol and ethanol oxidation [18,182,185].

In the particular case of ethanol, the production of carbon dioxide requires the breaking of the C−C bond of the ethanol molecule. This part of the mechanism of ethanol oxidation cannot be performed through a bifunctional mechanism and then, while PtRu is a very efficient catalyst for methanol oxidation, the use of a different cocatalyst is required for ethanol oxidation. The addition of tin to platinum has proven to increase the selectivity towards C−C bond breaking of ethanol molecule [18,186]. Furthermore, the addition of ruthenium and tin to platinum allows one to combine the catalytic effect of both metals: a bifunctional mechanism brought by ruthenium for water activation at low potential and the higher capacity of C−C bond breaking brought by tin [18,186]. The combination of these two effects allows increasing the oxidation current of ethanol at a given potential on PtRuSn alloys compared to pure Pt (Fig. 20) by increasing the number of electron exchanged per ethanol molecule oxidized (effect of the

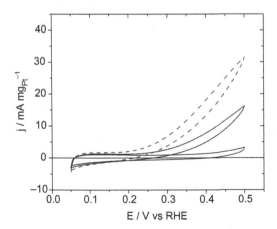

Figure 20 Cyclic voltammograms recorded in a 0.1 M $HClO_4$ + 0.1 M ethanol solution on different Pt-based electrodes with a Pt loading of 0.1 mg_{Pt} cm^{-2} at v = 5 mV s^{-1} and T = 20°C [18].

selectivity towards the formation of carbon dioxide) and by decreasing the surface poisoning by adsorbed CO (water activation at lower potential).

Since platinum is the base metal for these alloys, the catalytic materials are generally prepared through colloidal synthesis methods to obtain nanoparticles that are then dispersed on a carbon support (Fig. 21). The use of nanoparticles allows a maximization of the electrochemically active surface area of the catalytic material. The preparation of these nanomaterials is generally performed by colloidal synthesis methods at low temperature [122]. The strategy for colloidal synthesis methods (water-in-oil [187], nanocapsule [188], polyol [180], Bönnemann [189], etc.) is to dissolve the metal salt in an aqueous or organic medium containing a surfactant. The nanoparticles are then formed after a nucleation growth process occurs at low temperature, below 200°C generally, leading to a colloid (a stable suspension of solid particles in a solution). The support (carbon powder) is then dispersed in the colloid, and the catalyst is generally cleaned by heat treatment. These synthesis methods allow the preparation of nanoparticles with sizes in the range from 1 to 5 nm for Pt (Fig. 21A), PtRu (Fig. 21B), PtSn (Fig. 21E and F), or PtSnRu.

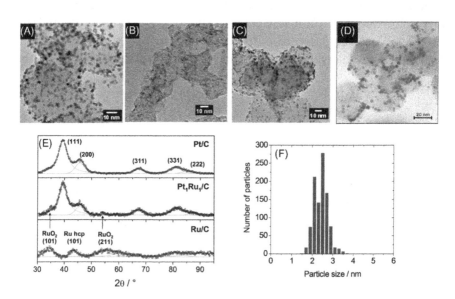

Figure 21 TEM images of Pt/C (A), PtRu/C (B), and Ru/C (C) catalysts prepared by colloidal synthesis with the corresponding XRD patterns (D) [180]. TEM image of Pt_9Sn_1/C catalyst prepared by colloidal synthesis method (E) with the corresponding size distribution (F) [18].

Due to the synthesis conditions, the formation of alloys by colloidal method is performed at very low temperature, and the alloying conditions may differ from those observed for bulk materials generally prepared at higher temperatures. Fig. 21A–D presents TEM images and the corresponding X-ray diffraction (XRD) pattern of Pt, PtRu, and Ru nanoparticles dispersed on a carbon support prepared for methanol oxidation [180]. The formation of a PtRu alloy was successful, and, generally, platinum-based catalysts adopt the face centered cubic (fcc) structure of platinum. However, the synthesis performed below 150°C did not allow a perfect alloying of the two metals, and some diffraction peaks from the RuO_2 rutile structure are visible on the XRD pattern of PtRu (Fig. 21D).

Recently, the development of ethanol electroreforming in alkaline medium has allowed the use of nonplatinum-based catalysts [173,190,191]. The activation of alcohol oxidation reaction occurs at lower potential, and catalyst displays a higher catalytic activity in alkaline medium than in acidic medium. Furthermore, the stability of nonplatinum metals such as palladium or nickel is greatly improved in alkaline medium compared to acidic medium. The selectivity of the alcohol oxidation reaction is then modified by the medium: in alkaline environment, the selectivity is oriented towards the formation of carboxylate ions. Hence, in alkaline medium, the increase of stability of nonplatinum metals and the increase of catalytic activity come at the expense of the amount of hydrogen produced per alcohol molecule consumed. This constraint led to the idea that the use of a valuable alcohol into platform molecules for further chemical transformation could be of interest in alkaline medium.

4.1.3.2 Ethylene Glycol and Glycerol

Among polyols, ethylene glycol and glycerol present the advantage to be simple molecules and can be produced from biomass. Furthermore, glycerol is a byproduct of biofuel industry, and its use for hydrogen production and potentially value-added compounds from glycerol electrooxidation would make the biofuel industry more profitable. Hence, the activation of their oxidation reactions can be performed more easily than complex polyols, especially in alkaline medium, and the resulting reaction products are interesting for further chemical modification for an application in the domains of pharmaceutical additives, surfactants, or polymers [192]. The reaction pathways of the oxidation reaction of

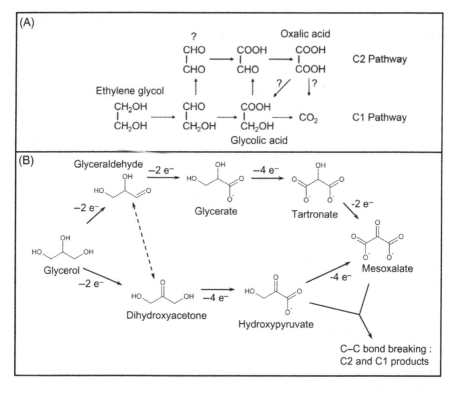

Figure 22 Possible reaction pathways for ethylene glycol oxidation reaction [194] (A) and glycerol oxidation reaction (B)

these two polyols described in Fig. 22 are complex and evidence the formation of several molecules of interest (aldehydes, ketones, and carboxylic acids) for chemistry, without C–C bond breaking. Their oxidation is then generally performed in alkaline medium with the aim of producing hydrogen and value-added coproducts [191–194]. In alkaline medium, the scope of catalytic metals available is considerably extended compared to acidic medium. Hence, the catalytic materials for ethylene glycol and glycerol oxidation reactions can vary from platinum- [19,172,195–197], palladium- [19,191,194,197–199], gold- [193,200,201], nickel- [202,203], or rhodium- [204] based catalysts. The nature of the metal tends to orientate the reaction pathway. In the case of ethylene glycol, both C1 and C2 pathways of Fig. 22A are observed with the formation of glycolate and oxalate ions on palladium- and platinum-based catalysts [172,191,194]. In the case of glycerol, platinum and palladium lead to the formation of reaction products present on the top pathway of Fig. 22B (glycerate ion, tartronate, and/or mesoxalate ions)

[19,191−193,196−199]. In the case of gold as catalyst, the same ions are detected, but hydroxypyruvate ion formation was also reported, indicating that gold may allow a different pathway for glycerol oxidation (the bottom pathway on Fig. 22B) [193,200,201]. For these three catalysts, if the potential is increased over 0.8 V vs RHE, the formation of CO_2 occurs indicating the destruction of the carbon chain for both ethylene glycol and glycerol. This C−C bond breaking is not suitable for the recovery of value-added molecules in the reaction products. In the case of Ni-based catalysts, the glycerol oxidation occurs after the $Ni^{II}(OH)_2 \rightarrow Ni^{III}OOH$ transition, i.e., above 1.0 V vs RHE. In this potential range, the main reaction products are glycerate ions with small amount of tartronate and oxalate ions, glycolate ions, formate ions, and finally carbon dioxide [203]. This product distribution indicates a rapid formation of glycerate ions acting as both reaction product and reaction intermediate for the formation of smaller molecules after C−C bond breaking.

The formation of PdAu alloys lead to decrease the onset potential of glycerol oxidation compared to that measured on Pd or Au catalysts [192,193,200]. This synergistic effect observed with these catalysts does not modify the global selectivity, with an initial oxidation of glycerol into glyceraldehyde as main reaction pathway (top pathway on Fig. 22B). However, the decrease of the oxidation potential allows a control of the selectivity by the anode operating potential. Indeed, at low potentials (below 0.8 V vs RHE), the presence of low oxidation reaction products is reported (mainly glycerate ions) while at higher potentials (over 0.8 V vs RHE), reaction products with a high oxidation degree are reported (hydroxypyruvate, tartronate, and mesoxalate ions) as well as carbon dioxide and carbonate ions, indicating a C−C bond breaking during the oxidation reaction [192,193,200].

In order to improve the catalytic activity of platinum and palladium for glycerol oxidation, bismuth can be added to the catalyst. PtBi as well as PdBi catalysts display a considerable increase of catalytic activity compared to Pt and Pd, but also a change in the reaction pathway [19,195−199]. The initial oxidation of the secondary alcohol group appears on both PtBi and PdBi catalysts and on PtPdBi catalysts. The formation of an alloy between Pt or Pd and Bi is not attested since no significant shift of the XRD peaks could be observed [193]. Furthermore, increasing the amount of bismuth over 10 at% in the catalytic materials leads to the formation of bismuth oxide structures.

Hence, bismuth seems to be dispersed as small atom clusters on the surface of platinum and palladium. Indeed, the formation of an alloy is not required to observe the increase of activity and the change of selectivity, and the same phenomenon can be observed from bismuth adatoms [181,195,199]. The change in glycerol oxidation reaction selectivity can then be explained by a change of glycerol adsorption mode due to the presence of bismuth adatoms hindering a part of the adsorption and reaction surface sites (Fig. 23). The associated increase of activity for PtBi and PdBi catalysts allows an increase of selectivity over the operating potential domain of the anode. Below 0.6 V *vs* RHE, aldehydes and ketones are detected with glycerate ions, in the potential range from 0.6 V to 0.8 V *vs* RHE, hydroxypyruvate, tatronate, and mesoxalate ions are detected while over 0.8 V *vs* RHE, the detection of carbon dioxide indicates a C−C bond breaking during the oxidation reaction.

The structure of the nanocatalyst can also induce a change in reaction selectivity. Generally, due to the presence of noble and expensive metals in the catalytic materials used for glycerol oxidation, they are present as nanoparticles dispersed on a support (Fig. 24A and B). Other structures such as controlled shape nanoparticles presenting a high surface orientation (Fig. 24C and D) and nanofoams (Fig. 24E and F) have been studied for glycerol oxidation.

Nanofoams are an excellent example of the effect of nanostructure on the selectivity of the catalyst: In the case of PdBi nanofoam, the same glycerol oxidation reaction pathway is observed as for nanoparticles dispersed on carbon, but the selectivity for the formation of

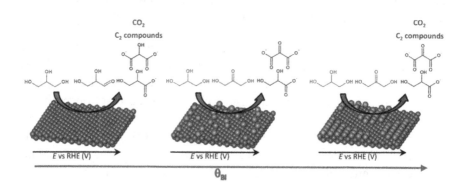

Figure 23 Change in selectivity of glycerol oxidation reaction induced by bismuth surface coverage (θ_{Bi}) [181].

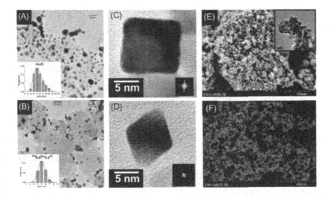

Figure 24 TEM images of Au/C (A) and $Pd_{0.5}Au_{0.5}/C$ (B) nanoparticles. Inset: Corresponding size distribution [193]. TEM images and corresponding FFT pattern of Pd nanocube (C) and Pd nanooctahedron (D) [181]. SEM image of PdBi (E) and PdSn nanofoams (F). Inset: TEM image of the nanoparticles constituting the nanofoam [205,206].

hydroxypyruvate is increased due to a longer contact time of the reactant trapped in the nanoporosity of the catalyst [205]. The same effect is observed with PdSn nanofoams with a glycerol oxidation reaction occurring mainly via glyceraldehyde as intermediate. Indeed, contrary to bismuth, the addition of tin to palladium does not change the reaction pathway. With a PdSn nanofoam, the presence of nanoporous cavities inside the catalytic material leads to a selectivity towards the exclusive formation of carboxylates [206].

4.1.4 Cell Performances

For alcohol electroreforming cell, the notion of performance depends on the goals to meet. For simple alcohols such as ethanol and methanol, the main objective is to produce hydrogen at low cost while consuming the alcohol into carbon dioxide. Reaching a high current density at low cell voltage (below 1.0 V) with selectivity towards carbon dioxide is then the aim of a methanol/ethanol electroreforming cell, and it is generally performed in PEMEC (i.e., in acidic medium).

In the case of methanol, the activation of the oxidation reaction can be performed at low potential and the production of byproducts such as formaldehyde and formic acid remains limited. Hence, the selectivity along the whole operating potential range is oriented towards the formation of carbon dioxide [176,183]. Current densities of PEMEC operating with methanol containing solution on a Pt or a PtRu anode can raise up to 1.5 A cm^{-2} at 1.0 V and 80°C [176,183].

78   Hydrogen Electrochemical Production

In the case of ethanol, the mechanism of the oxidation reaction into carbon dioxide involves more electrochemical steps, and the reaction then becomes more difficult to activate than methanol oxidation. Lamy et al. performed the electroreforming of ethanol in a proton exchange membrane electrolysis cell with analysis of reaction products by HPLC (Fig. 25A) [18]. The system was operated at 20°C and could reach $0.12\ A\ cm^{-2}$ at $1.0\ V$ (Fig. 25B). Using a similar setup, Caravaca et al. could increase the current density up to $0.2\ A\ cm^{-2}$ by increasing the temperature to 80°C [207]. The chemical yield of ethanol oxidation in this electroreforming cell was assessed by the analysis of reaction product during stationary electrolysis at $0.1\ A\ cm^{-2}$ and 20°C

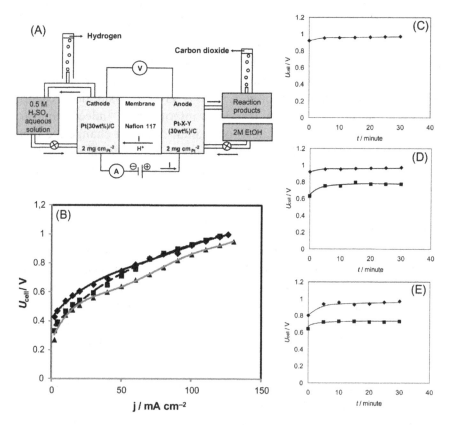

Figure 25 Schematic principle of the electrochemical decomposition of water and ethanol in a proton exchange membrane electrolysis cell (A), electrolysis cell voltage versus current density, $U_{cell}(j)$, for the electroreforming of ethanol (0.5 M H_2SO_4, Pt/C, Nafion 117, $Pt_xSn_yRu_z$/C, 2 M CH_3CH_2OH, $V_{initial}$ = 1 L, flow rate = 2 mL min^{-1}) at 20°C, ♦ Pt/C anode, ■ $Pt_{90}Sn_{10}$/C anode, ▲ $Pt_{86}Sn_{10}Ru_4$/C (B), electrolysis cell voltage versus time, $U_{cell}(t)$, at 20°C with a Pt/C anode (C), a $Pt_{90}Sn_{10}$/C anode (D), and a $Pt_{86}Sn_{10}Ru_4$/C anode (E). A volume of 2-M ethanol, $V_{initial}$ = 1 L, flow rate = 2 mL min^{-1}; ♦ j = 100 mA cm^{-2}, and ▲ j = 50 mA cm^{-2} [18].

with three different anodes (Pt, $Pt_{90}Sn_{10}/C$, and $Pt_{86}Sn_{10}Ru_4/C$). The results are presented in Table 5.

By increasing the selectivity towards the formation of carbon dioxide, $Pt_{90}Sn_{10}/C$ and $Pt_{86}Sn_{10}Ru_4/C$ increase the number of electrons exchanged per ethanol molecule oxidized. This parameter is of paramount importance in order to compare the performance of ethanol electrolysis cell since it will determine the amount of hydrogen molecules produced per ethanol molecule consumed and will have a strong impact on the final hydrogen cost. From the number of exchanged electrons for these two catalysts, an average of 2.3 molecules of H_2 is recovered per ethanol molecule consumed at 20°C. It represents more than one-third of the maximum theoretical maximum number of H_2 molecules recoverable from ethanol (Table 1).

The problematic of polyol electroreforming is slightly different: The aims are multiple with the production of hydrogen at low potential (below 1.0 V), high rate (high current density), and with the recovery of valuable coproducts from polyol oxidation reaction (no C−C bond breaking).

Gonzales-Cobos et al. performed the electroreforming of glycerol in alkaline medium on PtBi anode catalyst in a 5-cm^2 electrolysis cell [197]. The current densities of the system were close to 0.03 A cm^2 at 0.55 V and 0.06 A cm^{-2} at 0.7 V at 20°C (Fig. 26A). In order to avoid a C−C bond breaking observed over 0.8 V on this catalyst, the cell voltage was not increased above 0.7 V. Indeed, in the case of this study, the important parameter was the recovery of valuable coproducts with hydrogen such as glyceraldehyde, dihydroxyacetone, glycerate ions, or hydroxypyruvate

Table 5 Chemical Yield and Number of Electrons Exchanged Per Mole of Ethanol for the Electrolysis of 2 M Ethanol at $j = 0.1$ A cm^{-2} [18]						
Catalyst			Acetaldehyde	Acetic acid	CO_2	Calculated n_e^a
Pt/C	U_{cell}/V		0.92−0.97			3.5
	Chemical yield/mol%		46	42	12	
$Pt_{0.9}Sn_{0.1}/C$	U_{cell}/V		0.92−0.97			4.6
	Chemical yield/mol%		31	41	28	
$Pt_{0.86}Sn_{0.1}Ru_{0.04}/C$	U_{cell}/V		0.80−0.96			4.5
	Chemical yield/mol%		39	32	29	
^{a}Total number of mole of electrons exchanged for the oxidation of 1 mol of ethanol, assuming that no other electrolysis products than acetaldehyde, acetic acid, and CO_2 are formed.						

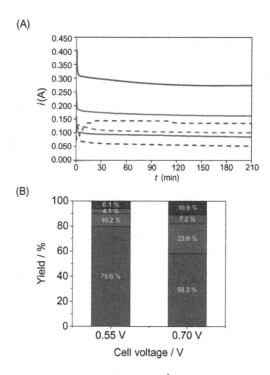

Figure 26 Electrolysis cell current versus time at 20°C for a 5 cm² electrolysis cell fitted with a Pt/C cathode and a Pt₉Bi₁/C anode and fed with 2 M glycerol + 0.5 M NaOH (red lines in online version), 2 M glycerol + 1.0 M NaOH (blue lines in online version), 2 M glycerol + 2.0 M NaOH (green lines in online version). Flow rate = 2 mL min⁻¹; U_cell = 0.55 V (dashed lines) and 0.70 V (plain lines) (A), product distribution analyzed by HPLC after 4 h electrolysis on a Pt₉Bi₁/C anode and fed with 2 M glycerol + 0.5 M NaOH at U_cell = 0.55 V and 0.70 V (B).

ions. The product distribution at these two cell voltages reveals the presence of glyceraldehyde in high extend at 0.55 V and its transformation into carboxylate ions at 0.70 V (Fig. 26B). This product distribution does not take dihydroxyacetone into account since it is not a stable molecule in aqueous alkaline medium, and it may only be a reaction intermediate for hydroxypyruvate ion formation.

Chen et al. obtained similar results for glycerol oxidation at low temperature (25°C) on Pd-based anode in alkaline medium [173]. The increase of temperature of their electroreforming system led to a global increase of the current densities over the whole potential range studied, reaching up to 1 A cm⁻² at a cell voltage of 0.9 V and at 80°C. However, the changes in selectivity induced by the temperature increase were not tracked, and hence the electroreforming system was operating under high hydrogen production rate conditions; it is not possible to establish if the reaction products from glycerol oxidation were still valuable coproducts.

4.2 SACCHARIDE REFORMING

The use of saccharides for hydrogen production would represent a more convenient way of introducing biosourced molecules in the hydrogen production chain. Indeed, a large part of cellulosic biomass that is not devoted to food industry is composed of saccharide derivatives (Fig. 3). The production of saccharides from these resources is then less demanding in terms of energy and chemical steps than alcohol/polyol production. Furthermore, the oxidation of saccharides in alkaline medium under conditions similar to that of an electroreforming cell [208] leads to the formation of carboxylic acids very similar to those produced from alcohol/polyol electroreforming and to longer carboxylic acids (Fig. 27).

Hence, a saccharide electroreforming process with a high control of anode selectivity could replace the saccharide treatment step into alcohol by directly coproducing the molecules of interest for chemical industries with hydrogen. However, before such electroreforming

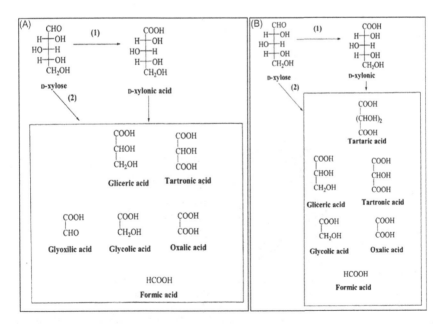

Figure 27 General mechanism for D-xylose oxidation in alkaline medium on Pt (A) and Au (B) [208].

system exists, it is necessary to improve the activity of catalysts used for the oxidation of saccharides. Indeed, the onset potentials of glucose (Fig. 28) and xylose oxidation reactions in alkaline medium are low compared to those of alcohols, and the reaction occurs on a wide potential range even on nonplatinum noble metals such as gold and palladium. However, the current densities measured are very low ($2.0 \, mA \, cm^{-2}$, Fig. 28A) compared to alcohol oxidation reaction performed on similar catalytic materials. The in situ analysis of reaction products reveals the formation of different carboxylic acids (evidenced by the absorption band at $1580 \, cm^{-1}$, Fig. 28B and C) over the whole potential range without poisoning of the surface by carbon monoxide.

The saccharide electroreforming is then an attractive way for producing hydrogen from renewable resources and sustainable conversion methods, which can benefit from the different development of alcohol electroreforming, but the arising of an efficient and performant system will require long research and development processes.

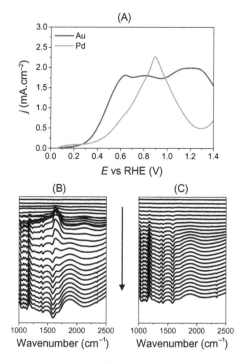

Figure 28 Linear voltammetry performed at $5 \, mV \, s^{-1}$ in 0.1 M glucose + 0.1 M NaOH electrolyte on Au/C (purple in online version) and Pd/C (green in online version) electrode (A). Infrared spectra recorded from 0.05 to 1.40 V vs RHE during glucose electrooxidation in 0.1 M glucose + 0.1 M NaOH electrolyte on Au/C (B) and Pd/C (C) electrodes.

CONCLUSION

Owing to the human population growth and the rapid development of emergent countries, the energy and goods demands will increase in the future. In this context, the only use of fossil resources is obviously non-sustainable for different reasons: their reserves will be exhausted very rapidly, and their burning produces nonenvironmental gases that have a strong impact on the global climate change and on health.

In order to address these challenges, hydrogen is considered the most significant candidate in technology innovation, economic expansion, and global progression in the 21st century. But hydrogen has to be produced. Currently, the reforming of fossil resources, and mainly of methane, is by far the most used method for the production of hydrogen. The reason is that the cost of hydrogen production from methane reforming is lower than that from electrolysis technology. Indeed, the cost of hydrogen from methane reforming is ca. $1.5 \, € \, kg_{H_2}^{-1}$. Now, according to Eq. (80), the electric energy consumed by a water electrolysis cell working at 1.8 V is $4.3 \, kW \, h \, N \, m_{H_2}^{-3}$, and considering an average price for electric energy of $0.15 \, € \, kW \, h^{-1}$, the cost of hydrogen is $0.645 \, € \, N \, m_{H_2}^{-3}$, c. $7.2 \, € \, kg_{H_2}^{-1}$. This value is in agreement with those given by Turner [209]. The cost estimated for hydrogen production method from fossil resources steam reforming does not take into account the environmental, societal, social, and health costs that will inescapably increase in the future, if the same rhythm of fossil fuels consumption is maintained, or worst, increases. The use of biogas for hydrogen production could be an alternative to natural gas, because it comes from renewable resources, and the CO_2 produced belongs to the natural carbon cycle. But the cleaning processes for obtaining high-purity hydrogen for chemical industry and energy conversion will remain very complex. Moreover, it has been shown that dependently on the renewable energy sources, especially hydro and wind energy, the cost of hydrogen from water electrolysis can be decreased [210]. But, important works have to be done in order to improve the water electrolysis cell efficiency, to decrease the capital cost of the plant (costs of materials, components, and systems), to increase the lifetime of the cell (components), and to decrease the maintenance requirements for the systems [210].

Another good alternative is the electroreforming of renewable compounds issued from biomass or wastes from biofuel industries, such as polyols and saccharides. Indeed the electrochemical reforming of such compounds with hydrogen production can be performed at very low cell voltages, lower than 1.0 V. This means that the electrical energy needed for the production of hydrogen can be more than twice lower than for water electrolysis. For example, the electrooxidation of glycerol can be performed in an electrolysis cell at 0.6 V, which corresponds to an electric energy consumption of 1.43 kW h N $m_{H_2}^{-3}$, i.e., 2.4 € $kg_{H_2}^{-1}$. This cost remains higher than that for the production of hydrogen from natural gas steam reforming, but the electrooxidation of glycerol can be controlled in such a way that compounds with very high value-added or industrial interest can be produced at the anode of the electrochemical cell simultaneously with the production of hydrogen at the cathode of the cell, which leads to decrease the production cost of hydrogen. Moreover, when the process is applied to for the conversion of wastes issued from agriculture or biofuel industries, it could help to make these economic sectors more profitable.

Indeed the European Commission and the DOE from the United States have very ambitious objectives concerning the biofuel utilization in the future. For example, the European Commission in 2007 proposed an "integrated energy and climate change" package, where 10% binding minimum for biofuels is targeted in 2020. The United States has the same objectives. Today the global biofuels market consists of approximately 85% bioethanol and 15% biodiesel. Bioethanol is produced and consumed mainly in Brazil and North America, whereas Europe is the world leader in biodiesel production. The increasing demand of methyl esters and bioethanol as fuels or fuel additives will then lead to the increase of crude glycerol stock (glycerol represents 10 and 4 wt% of products in biodiesel and bioethanol production, respectively), making this waste a cheaper raw material from chemistry. An effective usage or conversion of crude glycerol to specific products will then cut down the biodiesel production costs [211].

In conclusion the implementation of integrated systems based on the use of renewable energy (solar, wind, tide, etc.) for supplying electric energy to an electro-reforming cell working with valuable products from biomass, agriculture of biofuel industry wastes for the coproduction of hydrogen, and value added chemicals could be an excellent way to develop a sustainable economy respecting most of the "green chemistry" principles proposed by Anastas and Wagner [212], and to fit the definition of sustainable development given in the Brundtland Report [213].

ACKNOWLEDGEMENTS

The authors greatly acknowledge the CNRS Research Grouping HySPaC (GDR n°3652) and the Institute of Chemistry of the "Centre National de la Recherche Scientifique" (CNRS) for supporting this work.

REFERENCES

[1] P. Rullo, L. Nieto Degliuomini, M. Garciá, M. Basualdo, Model predictive control to ensure high quality hydrogen production for fuel cells, Int. J. Hydrogen Energy 39 (2014) 8635–8649.

[2] H. José Alves, C. Bley Jr, R. Rick Niklevicz, E. Pires Frigo, M. Sato Frigo, C.H. Coimbra-Araújo, Overview of hydrogen production technologies from biogas and the applications in fuel cells, Int. J. Hyd. Energy 38 (2013) 5215–5225.

[3] C.M. Kalamaras, A.M. Efstathiou, Hydrogen production technologies: current state and future developments, Conf. Papers Energy 2013 (2013) 1–9. Available from: http://dx.doi.org/10.1155/2013/690627.

[4] AFHYPAC (Association française pour l'hydrogène et les piles à combustible)—Production et consommation d'hydrogène aujourd'hui. Mémento de l'hydrogène. Fiche 1.3. http://afhypac.org/documents/tout.savoir/Fiche_1.3_revision_fevrier_2008.pdf.

[5] S.K. Mazloomi, N. Sulaima, Influencing factors of water electrolysis electrical efficiency, Renew. Sustain. Energy Rev. 16 (2012) 4257–4263.

[6] C. Coutanceau, S. Baranton, Electrochemical conversion of alcohols for hydrogen production: a short overview, WIRE Energy Environ. 5 (2016) 388–400.

[7] S. Balasubramanian, C.E. Holland, J.W. Weidner, Electrochemical removal of carbon monoxide in reformate hydrogen for fuelling proton exchange membrane fuel cells, Electrochem. Solid State Lett. 13 (2010) B5–B7.

[8] T.W. Napporn, J.M. Léger, C. Lamy, Electrocatalytic oxidation of carbon monoxide at lower potentials on platinum-based alloys incorporated in polyaniline, J. Electroanal. Chem. 408 (1996) 141–147.

[9] H.F. Oetjen, V.M. Schmidt, U. Stimming, F. Trila, Performance data of a proton exchange membrane fuel cell using H_2/CO as fuel gas, J. Electrochem. Soc. 143 (1996) 3838–3842.

[10] A. Marshall, Electrocatalysts for the oxygen evolution electrode in water electrolysers using proton exchange membranes: synthesis and characterization, Department of Materials Technology, Norwegian University of Science and Technology (NTNU), Trondheim, 2005.

[11] J.D. Holladay, J. Hu, D.L. King, Y. Wang, An overview of hydrogen production technologies, Catal. Today 139 (2009) 244–260.

[12] S. Baranton, C. Coutanceau, Nickel cobalt hydroxide nanoflakes as catalysts for the hydrogen evolution reaction, Appl. Catal. B: Environ. 136–137 (2013) 1–8.

[13] M.A. Dasari, P.P. Kiatsimkul, W.R. Sutterlin, G.J. Suppes, Low-pressure hydrogenolysis of glycerol to propylene glycol, Appl. Catal. A 281 (2005) 225–231.

[14] A.S. Aldiguier, S. Alfenore, X. Cameleyre, G. Goma, J.L. Uribelarrea, S.E. Guillouet, et al., Synergistic temperature and ethanol effect on Saccharomyces cerevisiae dynamic behaviour in ethanol bio-fuel production, Bioprocess Biosyst. Eng. 26 (2004) 217–222.

[15] R.H. Perry, D.W. Green (Eds.), Perry's Chemical Engineers Handbook, McGraw Hill, New York, 1999.

[16] Haynes W.M., Bruno T.J., Lide D.R., CRC Handbook of Chemistry and Physics, 97th ed., (2016).

[17] B. Jeffries-Nakamura, S.R. Narayanan, T.I. Valdez, W. Chun, Making hydrogen by electrolysis of methanol, NASA Tech Brief 6 (1948) 26.

[18] C. Lamy, T. Jaubert, S. Baranton, C. Coutanceau, Clean hydrogen generation through the electrocatalytic oxidation of ethanol in a proton exchange membrane electrolysis cell (PEMEC). Effect of the nature and structure of the catalytic anode, J Power Sources 245 (2014) 927–936.

[19] M. Simões, S. Baranton, C. Coutanceau, Enhancement of catalytic properties for glycerol electrooxidation on Pt and Pd nanoparticles induced by Bi surface modification, Appl. Catal. B: Environ. 110 (2011) 40–49.

[20] C. Bianchini, P.K. Shen, Palladium-based electrocatalysts for alcohol oxidation in half cells and in direct alcohol fuel cells, Chem. Rev. 109 (2009) 4183–4206.

[21] P. Panagiotopoulou, D.I. Kondarides, X.E. Verykios, Selective methanation of CO over supported noble metal catalysts: Effects of the nature of the metallic phase on catalytic performance, Appl. Catal. A: Gen. 344 (2008) 45–54.

[22] https://en.wikipedia.org/wiki/Hemicellulose#/media/File:Hemicellulose.png (accessed on March 29th, 2017).

[23] https://fr.wikipedia.org/wiki/Lignine#/media/File:LigninStructure.png (accessed on March 29th, 2017).

[24] NEUR Otiker—Own work, Public Domain, https://commons.wikimedia.org/w/index.php? curid = 3962569 (accessed on March 29th, 2017).

[25] By Vaccinationist. Own work, based on PubChem, Public Domain, https://commons.wikimedia.org/w/index.php?curid = 53307449 (accessed on March 29th, 2017).

[26] M. Ni, D.Y.C. Leung, M.K.H. Leung, A review on reforming bio-ethanol for hydrogen production, Int. J. Hydrogen Energy 32 (2007) 3238–3247.

[27] S.N. Naik, V.V. Goud, P.K. Rout, A.K. Dalai, Production of first and second generation biofuels: a comprehensive review, Renew. Sustain. En. Rev. 14 (2010) 578–597.

[28] .Dupré D., Bion N., Epron F., Production d'hydrogène par reformage du bioéthanol, Techniques de l'ingénieur, (2014).

[29] F. Auprêtre, C. Descormes, D. Duprez, Bio-ethanol catalytic steam reforming over supported metal catalysts, Catal. Comm. 3 (2002) 263–267.

[30] T. Montini, L. De Rogatis, V. Gombac, P. Fornasiero, M. Graziani, Rh(1%) @$Ce_xZr_{1-x}O_2-Al_2O_3$ nanocomposites: active and stable catalysts for ethanol steam reforming, Appl. Catal. B: Env. 71 (2007) 125–134.

[31] H.–S. Roh, Y. Wang, D.L. King, Selective production of H_2 from ethanol at low temperatures over Rh/ZrO_2-CeO_2 catalysts, Top. Catal. 49 (2008) 32–37.

[32] A. Le Valant, A. Garron, N. Bion, F. Epron, D. Duprez, Hydrogen production from raw bioethanol over $Rh/MgAl_2O_4$ catalyst. Impact of impurities: heavy alcohol, aldehyde, ester, acid and amine, Catal. Today 138 (2008) 169–174.

[33] A. Le Valant, F. Can, N. Bion, D. Duprez, F. Epron, Hydrogen production from raw bioethanol steam reforming: optimization of catalyst composition with improved stability against various impurities, Int. J. Hyd. Energy 35 (2010) 5015–5020.

[34] F. Mueller-Langer, E. Tzimas, M. Kaltschmitt, S. Peteves, Techno-economic assessment of hydrogen production processes for the hydrogen economy for the short and medium term, Int. J. Hyd. Energy 32 (2007) 3797–3810.

[35] A. Goñi-Urtiaga, D. Presvytes, K. Scott, Solid acids as electrolyte materials for proton exchange membrane (PEM) electrolysis: review, Int. J. Hydrogen Energy 37 (2012) 3358–3372.

[36] Millet P., Électrolyseurs de l'eau à membrane acide, Techniques de l'ingénieur, (2007)

[37] C. Lamy, From hydrogen production by water electrolysis to its utilization in a PEM fuel cell or in a so fuel cell: some considerations on the energy efficiencies, Int. J. Hydrogen Energy 41 (2016) 15415−15425.

[38] M.H. Miles, M.A. Thomason, Periodic variations of overvoltages for water electrolysis in acid solutions from cyclic voltammetric studies, J. Electrochem. Soc. 123 (1976) 1459−1461.

[39] M.H. Miles, E.A. Klaus, B.P. Gunn, J.R. Locker, W.E. Serafin, S. Srinivasan, The oxygen evolution reaction on platinum, iridium, ruthenium and their alloys at 80°C in acid solutions, Electrochim. Acta 23 (1978) 521−526.

[40] S. Trasatti, Electrocatalysis in the anodic evolution of oxygen and chlorine, Electrochim. Acta 29 (1984) 1503−1512.

[41] G. Lodi, E. Sivieri, A. Battisti, S. Trasatti, Ruthenium dioxide-based film electrodes: III. Effect of chemical composition and surface morphology on oxygen evolution in acid solutions, J. Appl. Electrochem. 8 (1978) 135−143.

[42] L.M. Doubova, A. De Battisti, S. Daolio, C. Pagura, S. Barison, R. Gerbasi, et al., Effect of surface structure on behavior of RuO_2 electrodes in sulfuric acid aqueous solution, Russ. J. Electrochem. 40 (2004) 1115−1122.

[43] Y. Matsumoto, E. Sato, Electrocatalytic properties of transition metal oxides for oxygen evolution reaction, Mater. Chem. Phys. 14 (1986) 397−426.

[44] D.B. Rogers, R.D. Shannon, A.W. Sleight, J.L. Gillson, Crystal chemistry of metal dioxides with rutile-related structures, Inorg. Chem. 8 (1969) 841−849.

[45] C. Iwakura, K. Hirao, H. Tamura, Anodic evolution of oxygen on ruthenium in acidic solutions, Electrochim. Acta 22 (1977) 329−334.

[46] R. Kötz, S. Stucki, Stabilisation of RuO_2 by IrO_2 for anodic oxygen evolution in acid media, Electrochim. Acta 31 (1986) 1311−1316.

[47] R. Kötz, RuO_2/IrO_2 electrocatalysts for anodic O_2 evolution, Electrochim. Acta 29 (1984) 1607−1612.

[48] R. Kötz, S. Stucki, D. Scherson, D.M. Kolb, In-situ identification of RuO_4 as the corrosion product during oxygen evolution on ruthenium in acid media, J. Electroanal. Chem. 172 (1984) 211−219.

[49] A. Minguzzi, F.R.F. Fan, A. Vertova, S. Rondinini, A.J. Bard, Dynamic potential−Ph diagrams application to electrocatalysts for water oxidation, Chem. Sci. 3 (2012) 217.

[50] W. Xu, S. Keith, RuO_2 supported on Sb-doped SnO_2 nanoparticles for polymer electrolyte membrane water electrolysers, Int. J. Hydrogen Energy 36 (2011) 5806−5810.

[51] J.C. Cruz, S. Rivas, D. Beltran, Y. Meas, R. Ornelas, G. Osorio.Monreal, et al., Synthesis and evaluation of ATO as a support for Pt.IrO_2 in a unitized regenerative fuel cell, Int. J. Hydrogen Energy 37 (2012) 13522−13528.

[52] S. Siracusano, V. Baglio, C. D'Urso, V. Antonucci, A.S. Arico, Preparation and characterization of titanium suboxides as conductive supports of IrO_2 electrocatalysts for application in SPE electrolysers, Electrochim. Acta 54 (2009) 6292−6299.

[53] L. Ma, S. Sui, Y. Zhai, Preparation and characterization of Ir/TiC catalyst for oxygen evolution, J. Power Sources 177 (2008) 470−477.

[54] S.-Y. Huang, P. Ganesan, H.-Y. Jung, B.N. Popov, Development of supported bifunctional oxygen electrocatalysts and corrosion-resistant gas diffusion layer for unitized regenerative fuel cell applications, J. Power Sources 198 (2012) 23−29.

[55] G. García, M. Roca-Ayats, A. Lillo, J.L. Galante, M.A. Pena, M.V. Martínez-Huerta, Catalyst support effects at the oxygen electrode of unitized regenerative fuel cells, Catal. Today 210 (2013) 67−74.

[56] S. Trasatti, Physical electrochemistry of ceramic oxides, Electrochim. Acta 36 (1991) 225–241.

[57] J.C. Cruz, V. Baglio, S. Siracusano, V. Antonucci, A.S. Arico, R. Ornelas, et al., Preparation and characterization of RuO$_2$ catalysts for oxygen evolution in a solid polymer electrolyte, Int. J. Electrochem. Sci. 6 (2011) 6607–6619.

[58] J.P. Zheng, A new charge storage mechanism for electrochemical capacitors, J. Electrochem. Soc. 142 (1995) L6.

[59] A. Devadas, S. Baranton, T.W. Napporn, C. Coutanceau, Tailoring of RuO$_2$ nanoparticles by microwave assisted "Instant method" for energy storage applications, J. Power Sources 196 (2011) 4044–4053.

[60] E. Tsuji, A. Imanishi, K.I. Fukui, Y. Nakato, Electrocatalytic activity of amorphous RuO$_2$ electrode for oxygen evolution in an aqueous solution, Electrochim. Acta 56 (2011) 2009–2016.

[61] C. Sassoye, G. Muller, D.P. Debecker, A. Karelovic, S. Cassaignon, C. Pizarro, et al., A sustainable aqueous route to highly stable suspensions of monodispersed nano ruthenia, Green Chem. 13 (2011) 3230.

[62] H. Kim, B.N. Popov, Characterization of hydrous ruthenium oxide/carbon nanocomposite supercapacitors prepared by a colloidal method, J. Power Sources 104 (2002) 52–61.

[63] Y. Murakami, S. Ichikawa, Y. Takasu, Preparation of ultrafine ruthenium oxide particles with ammonium hydrogencarbonate, Electrochemistry 65 (1997) 992–996.

[64] H. Ma, C. Liu, J. Liao, Y. Su, X. Xue, W. Xing, Study of ruthenium oxide catalyst for electrocatalytic performance in oxygen evolution, J. Mol. Catal. A: Chem. 247 (2006) 7–13.

[65] J. Cheng, H. Zhang, G. Chen, Y. Zhang, Study of Ir$_x$Ru$_{1-x}$O$_2$ oxides as anodic electrocatalysts for solid polymer electrolyte water electrolysis, Electrochim. Acta 54 (2009) 6250–6256.

[66] A. Marshall, B. Børresen, G. Hagen, M. Tsypkin, R. Tunold, Preparation and characterisation of nanocrystalline Ir$_x$Sn$_{1-x}$O$_2$ electrocatalytic powders, Mater. Chem. Phys. 94 (2005) 226–232.

[67] X. Wu, J. Tayal, S. Basu, K. Scott, Nano-crystalline Ru$_x$Sn$_{1-x}$O$_2$ powder catalysts for oxygen evolution reaction in proton exchange membrane water electrolysers, Int. J. Hydrogen Energy 36 (2011) 14796–14804.

[68] J. Wen, X. Ruan, Z. Zhou, Preparation and electrochemical performance of novel ruthenium–manganese oxide electrode materials for electrochemical capacitors, J. Phys. Chem. Solids 70 (2009) 816–820.

[69] A. Di Blasi, C. D'Urso, V. Baglio, V. Antonucci, A.S. Arico, R. Ornelas, et al., Preparation and evaluation of RuO$_2$–IrO$_2$, IrO$_2$–Pt and IrO$_2$–Ta$_2$O$_5$ catalysts for the oxygen evolution reaction in an SPE electrolyzer, J. Appl. Electrochem. 39 (2008) 191–196.

[70] L.E. Owe, M. Tsypkin, K.S. Wallwork, R.G. Haverkamp, S. Sunde, Iridium–ruthenium single phase mixed oxides for oxygen evolution: composition dependence of electrocatalytic activity, Electrochim. Acta 70 (2012) 158–164.

[71] N. Mamaca, E. Mayousse, S. Arrii-Clacens, T.W. Napporn, K. Servat, N. Guillet, et al., Electrochemical activity of ruthenium and iridium based catalysts for oxygen evolution reaction, Appl. Catal. B: Environ. 111–112 (2012) 376–380.

[72] J.M. Roller, M.J. Arellano-Jimenez, R. Jain, H. Yu, C. Barry Carter, R. Maric, Oxygen evolution during water electrolysis from thin films using bimetallic oxides of Ir–Pt and Ir–Ru, J. Electrochem. Soc. 160 (2013) F716–F730.

[73] L.A. De Faria, S. Trasatti, Effect of composition on the point of zero charge of RuO$_2$ + TiO$_2$ mixed oxides, J. Electroanal. Chem. 340 (1992) 145–152.

[74] T. Audichon, E. Mayousse, S. Morisset, C. Morais, C. Comminges, T.W. Napporn, et al., Electroactivity of RuO$_2$–IrO$_2$ mixed nanocatalysts toward the oxygen evolution reaction in a water electrolyzer supplied by a solar profile, Int. J. Hydrogen Energy 39 (2014) 16785–16796.

[75] T. Audichon, B. Guenot, S. Baranton, M. Cretin, C. Lamy, C. Coutanceau, Preparation and characterization of supported Ru$_x$Ir$_{(1-x)}$O$_2$ nano-oxides using a modified polyol synthesis assisted by microwave activation for energy storage applications, Appl. Catal. B: Environ. 200 (2017) 493–502.

[76] T. Audichon, T.W. Napporn, C. Canaff, C. Morais, C. Comminges, K.B. Kokoh, IrO$_2$ coated on RuO$_2$ as efficient and stable electroactive nanocatalysts for electrochemical water splitting, J. Phys. Chem. C. 120 (2016) 2562–2573.

[77] J. Xu, G. Liu, J. Li, X. Wang, The electrocatalytic properties of an IrO$_2$/SnO$_2$ catalyst using SnO$_2$ as a support and an assisting reagent for the oxygen evolution reaction, Electrochim. Acta 59 (2012) 105–112.

[78] W. Xu, K. Scott, S. Basu, Performance of a high temperature polymer electrolyte membrane water electrolyser, J. Power Sources 196 (2011) 8918–8924.

[79] J.L. Corona-Guinto, L. Cardeño-García, D.C. Martínez-Casillas, P. Sandoval-Pineda, J.M. Tamayo-Meza, R. Silva-Casarin, et al., Performance of a PEM electrolyzer using RuIrCoOx electrocatalysts for the oxygen evolution electrode, Int. J. Hydrogen Energy 38 (2013) 12667–12673.

[80] V. Baglio, A. Di Blasi, T. Denaro, V. Antonucci, A.S. Arico, R. Ornelas, et al., Arriaga, synthesis, characterization and evaluation of IrO$_2$–RuO$_2$ electrocatalytic powders for oxygen evolution reaction, J. New Mater. Electrochem. Syst. 11 (2008) 105–108.

[81] S. Song, H. Zhang, X. Ma, Z. Shao, R.T. Baker, B. Yi, Electrochemical investigation of electrocatalysts for the oxygen evolution reaction in PEM water electrolyzers, Int. J. Hydrogen Energy 33 (2008) 4955–4961.

[82] L.Å. Näslund, A.S. Ingason, S. Holmin, J. Rosen, Formation of RuO(OH)$_2$ on RuO$_2$-based electrodes for hydrogen production, J. Phys. Chem. C 118 (2014) 15315–15323.

[83] J.M. Kahk, et al., Understanding the electronic structure of IrO$_2$ using hard X-ray photoelectron spectroscopy and density. Functional theory, Phys. Rev. Lett. 112 (2014) 117601.

[84] M. Schaefer, R. Schlaf, Electronic structure investigation of atomic layer deposition ruthenium (oxide) thin films using photoemission spectroscopy, J. Appl. Phys. 118 (2015) 065306.1–065306.7.

[85] D. Rochefort, P. Dabo, D. Guay, P.M.A. Sherwood, XPS investigations of thermally prepared RuO$_2$ electrodes in reductive conditions, Electrochim. Acta 48 (2003) 4245–4252.

[86] M.P. Browne, H. Nolan, G.S. Duesberg, P.E. Colavita, M.E.G. Lyons, Low-overpotential high-activity mixed manganese and ruthenium oxide electrocatalysts for oxygen evolution reaction in alkaline media, ACS Catal. 6 (2016) 2408–2415.

[87] Y.S. Huang, P.C. Liao, Preparation and characterization of RuO$_2$ thin films, Sol. Energy Mat. Sol. C. 55 (1998) 179–197.

[88] O. Barbieri, M. Hahn, A. Foelske, R. Kotz, Effect of electronic resistance and water content on the performance of RuO$_2$ for supercapacitors, J. Electrochem. Soc. 153 (2006) A2049–A2054.

[89] S. Trasatti, G. Buzzanca, Ruthenium dioxide: a new interesting electrode material. Solid state structure and electrochemical behavior, J. Electroanal. Chem. Interfacial Electrochem. 29 (1971) A1–A5.

[90] F.I. Mattos-Costa, P. Lima-Neto, S.A.S. Machado, L.A. Avaca, Characterisation of surfaces modified by sol–gel derived Ru$_x$Ir$_{1-x}$O$_2$ coatings for oxygen evolution in acid medium, Electrochim. Acta 44 (1998) 1515–1523.

[91] V.A. Saveleva, L. Wang, W. Luo, S. Zafeiratos, C. Ulhaq-Bouillet, A.S. Gago, et al., Uncovering the stabilization mechanism in bimetallic ruthenium – iridium anodes for proton exchange membrane electrolyzers, J. Phys. Chem. Lett. 7 (2016) 3240–3245.

[92] K. Juodkazis, J. Juodkazyte, V. Sukiene, A. Griguceviciene, A. Selskis, On the charge storage mechanism at $RuO_2/0.5$ M H_2SO_4 interface, J. Solid State Electrochem. 12 (2008) 1399–1404.

[93] T. Audichon, E. Mayousse, T.W. Napporn, C. Morais, C. Comminges, K.B. Kokoh, Elaboration and characterization of ruthenium nano-oxides for the oxygen evolution reaction in a proton exchange membrane water electrolyzer supplied by a solar profile, Electrochim. Acta 132 (2014) 284–291.

[94] W. Sugimoto, T. Kizaki, K. Yokoshima, Y. Murakami, Y. Takasu, Evaluation of the pseudocapacitance in RuO_2 with a RuO_2/GC thin film electrode, Electrochim. Acta 49 (2004) 313–320.

[95] W. Sugimoto, K. Yokoshima, Y. Murakami, Y. Takasu, Charge storage mechanism of nanostructured anhydrous and hydrous ruthenium-based oxides, Electrochim. Acta 52 (2006) 1742–1748.

[96] K. Kuratani, T. Kiyobayashi, N. Kuriyama, Influence of the mesoporous structure on capacitance of the RuO_2 electrode, J. Power Sources 189 (2009) 1284–1291.

[97] J.W. Long, K.E. Swider, C.I. Merzbacher, D.R. Rolison, Voltammetric characterization of ruthenium oxide-based aerogels and other RuO_2 solids: the nature of capacitance in nanostructured materials, Langmuir 15 (1999) 780–785.

[98] S. Ardizzone, G. Fregonara, S. Trasatti, "Inner" and "Outer" active surface of RuO_2 electrodes, Electrochim. Acta 35 (1990) 263–267.

[99] H.S. Jeon, A.D. Chandra Permana, J. Kim, B.K. Min, Water splitting for hydrogen production using a high surface area RuO_2 electrocatalyst synthesized in supercritical water, Int. J. Hydrogen Energy 38 (2013) 6092–6096.

[100] C. Angelinetta, S. Trasatti, L.D. Atanasoska, R.T. Atanasoski, Surface properties of $RuO_2 + IrO_2$ mixed oxide electrodes, J. Electroanal. Chem. 214 (1986) 535–546.

[101] A. Marshall, B. Børresen, G. Hagen, S. Sunde, M. Tsypkin, R. Tunold, Iridium oxide. based nanocrystalline particles as oxygen evolution electrocatalysts, Russ. J. Electrochem. 42 (2006) 1134–1140.

[102] S. Siracusano, V. Baglio, A. Di Blasi, N. Briguglio, A. Stassi, R. Ornelas, et al., Electrochemical characterization of single cell and short stack PEM electrolyzers based on a nanosized IrO_2 anode electrocatalyst, Int. J. Hydrogen Energy 35 (2010) 5558–5568.

[103] J.J. Zhang, J.M. Hu, J.Q. Zhang, C.N. Cao, IrO_2-SiO_2 binary oxide films: geometric or kinetic interpretation of the improved electrocatalytic activity for the oxygen evolution reaction, Int. J. Hydrogen Energy 36 (2011) 5218–5226.

[104] L.A. De Faria, J.F.C. Boodts, S. Trasatti, Electrocatalytic properties of ternary oxide mixtures of composition $Ru_{0.3}Ti_{(0.7-x)}Ce_xO_2$: oxygen evolution from acidic solution, J. Appl. Electrochem. 26 (1996) 1195–1199.

[105] E. Antolini, Iridium as catalyst and co-catalyst for oxygen Evolution/Reduction in acidic polymer electrolyte membrane electrolyzers and fuel cells, ACS Catal. 4 (2014) 1426–1440.

[106] J.M. Hu, J.Q. Zhang, C.N. Cao, Oxygen evolution reaction on IrO_2-based DSA® type electrodes: kinetics analysis of Tafel lines and EIS, Int. J. Hydrogen Energy 29 (2004) 791–797.

[107] L.A. Da Silva, V.A. Alves, M.A.P. Da Silva, S. Trasatti, J.F.C. Boodts, Electrochemical impedance, SEM, EDX and voltammetric study of oxygen evolution on Ir + Ti + Pt ternary oxide electrodes in alkaline solution, Electrochim. Acta 41 (1996) 1279–1285.

[108] Y. Lai, Y. Li, L. Jiang, W. Xu, X. Lv, J. Li, et al., Electrochemical behaviors of co-deposited Pb/Pb.MnO$_2$ composite anode in sulfuric acid solution. Tafel and EIS investigations, J. Electroanal. Chem. 671 (2012) 16−23.

[109] F.R. Costa, D.V. Franco, L.M. Da Silva, Electrochemical impedance spectroscopy study of the oxygen evolution reaction on a gas-evolving anode composed of lead dioxide microfibers, Electrochim. Acta 90 (2013) 332−343.

[110] P. Brault, A. Caillard, A.L. Thomam, J. Mathias, C. Charles, R.W. Boswell, et al., Plasma sputtering deposition of platinum into porous fuel cell electrode, J. Phys. D: Appl. Phys. 37 (2004) 3419−3423.

[111] J. Perrière, E. Millon, M. Chamarro, M. Morcrette, C. Andreazza, Formation of GaAs nanocrystals by laser ablation, Appl. Phys. Lett 78 (2001) 2949.1−2949.3.

[112] E. Billy, F. Maillard, A. Morin, L. Guetaz, F. Emieux, C. Thurier, et al., Impact of ultra-low Pt loadings on the performance of anode/cathode in a proton-exchange membrane fuel cell, J. Power Sources 195 (2010) 2737−2746.

[113] C. Coutanceau, A. Rakotondrainibe, A. Lima, E. Garnier, S. Pronier, J.M. Léger, et al., Preparation of Pt−Ru bimetallic anodes by galvanostatic pulse electrodeposition: characterization and application to the direct methanol fuel cell, J. Appl. Electrochem. 34 (2004) 61−66.

[114] F. Vigier, C. Coutanceau, A. Perrard, E.M. Belgsir, C. Lamy, Development of anode catalyst for direct ethanol fuel cell, J. Appl. Electrochem. 34 (2004) 439−446.

[115] C. Coutanceau, S. Brimaud, L. Dubau, C. Lamy, J.M. Léger, S. Rousseau, et al., Review of different methods for developing nanoelelectrocatalysts for the oxidation of organic compounds, Electrochim. Acta 53 (2008) 6865−6880.

[116] R. Sellin, J.M. Clacens, C. Coutanceau, A thermogravimetric analysis/mass spectroscopy study of the thermal and chemical stability of carbon in the Pt/C catalyst system, Carbon 48 (2010) 2244−2254.

[117] S. Lankiang, M. Chiwata, S. Baranton, H. Uchida, C. Coutanceau, Oxygen reduction reaction at binary and ternary nanocatalysts based on Pt, Pd and Au, Electrochim. Acta 182 (2015) 131−142.

[118] S.A. Grigoriev, M.S. Mamat, K.A. Dzhus, G.S. Walker, P. Millet, Platinum and palladium nano-particles supported by graphitic nano-fibers as catalysts for PEM water electrolysis, Int. J. Hydrogen Energy 36 (2011) 4143−4147.

[119] D. Dru, S. Baranton, J. Bigarre, P. Buvat, C. Coutanceau, Fluorine-free Pt nanocomposites for three-phase interfaces in fuel cell electrodes, ACS Catal. 6 (2016) 6993−7001.

[120] C. Grolleau, C. Coutanceau, F. Pierre, J.M. Léger, Effect of potential cycling on structure and activity of Pt nanoparticles dispersed on different carbon supports, Electrochim. Acta 53 (2008) 7157−7165.

[121] S.A. Grigoriev, P. Millet, V.N. Fateev, Evaluation of carbon-supported Pt and Pd nano-particles for the hydrogen evolution reaction in PEM water electrolysers, J. Power Sources 177 (2008) 281−285.

[122] C. Coutanceau, S. Baranton, T.W. Napporn, Platinum fuel cell nanoparticle syntheses: effect on morphology, structure and electrocatalytic behavior in: A.Hashim(Ed.), The Delivery of Nanoparticles, InTech Publisher, Chapter. 19, Rijeka, 2011, pp. 403−430.

[123] J.G.N. Thomas, Kinetics of electrolytic hydrogen evolution and the adsorption of hydrogen by metals, Trans. Faraday Soc 57 (1961) 1603−1611.

[124] A.B. Laursen, A.S. Varela, F. Dionigi, H. Fanchiu, C. Miller, O.L. Trinhammer, et al., Electrochemical hydrogen evolution: Sabatier's principle and the volcano plot, J. Chem. Educ. 89 (2012) 1595−1599.

[125] J. Duan, S. Chen, M. Jaroniec, S.Z. Qiao, Heteroatom-doped graphene-based materials for energy-relevant electrocatalytic processes, ACS Catal. 5 (2015) 5207−5234.

[126] A. Zalineva, S. Baranton, C. Coutanceau, G. Jerkiewicz, Electrochemical behavior of unsupported shaped palladium nanoparticles, Langmuir 31 (2015) 1605−1609.

[127] P. Millet, F. Andolfatto, R. Durand, Design and performance of a solid polymer electrolyte water electrolyzer, Int. J. Hydrogen Energy 21 (1996) 87−93.

[128] A.S. Aricò, S. Siracusano, N. Briguglio, V. Baglio, A. Di Blasi, V. Antonucci, Polymer electrolyte membrane water electrolysis: status of technologies and potential applications in combination with renewable power sources, J. Appl. Electrochem. 43 (2013) 107−118.

[129] K.E. Ayers, E.B. Anderson, C.B. Capuano, B.D. Carter, L.T. Dalton, G. Hanlon, et al., Research advances towards low cost, high efficiency PEM electrolysis, ECS Trans. 33 (2010) 3−15.

[130] Y. Kawano, Y. Wang, R.A. Palmer, S.R. Aubuchon, Stress−strain curves of nafion membranes in acid and salt forms, Polim.: Cienc. Tecnol. 12 (2002) 96−101.

[131] S.A. Grigoriev, V.I. Porembsky, V.N. Fateev, Pure hydrogen production by PEM electrolysis for hydrogen energy, Int. J. Hydrogen Energy 31 (2006) 171−175.

[132] H. Ito, T. Maeda, A. Nakano, H. Takenaka, Properties of nafion membranes under PEM water electrolysis conditions, Int. J. Hydrogen Energy 36 (2011) 10527−10540.

[133] V. Antonucci, A. Di Blasi, V. Baglio, R. Ornelas, F. Matteucci, J. Ledesma-Garcia, et al., High temperature operation of a composite membrane-based solid polymer electrolyte water electrolyser, Electrochim. Acta 53 (2008) 7350−7356.

[134] S. Siracusano, V. Baglio, M.A. Navarra, S. Panero, V. Antonucci, A.S. Arico, Investigation of composite nafion/sulfated zirconia membrane for solid polymer electrolyte electrolyzer applications, Int. J. Electrochem. Sci. 7 (2012) 1532−1542.

[135] A. Albert, A.O. Barnett, M.S. Thomassen, T.J. Schmidt, L. Gubler, Radiation-grafted polymer electrolyte membranes for water electrolysis cells: evaluation of key membrane properties, ACS Appl. Mater. Interfaces 7 (2015) 22203−22212.

[136] Y. Wang, D.Y.C. Leung, J. Xuan, H. Wang, A review on unitized regenerative fuel cell technologies, part. A: Unitized regenerative proton exchange membrane fuel cells, Renew. Sustain. Energy Rev. 65 (2016) 961−977.

[137] C. Rozain, E. Mayousse, N. Guillet, P. Millet, Influence of iridium oxide loadings on the performance of PEM water electrolysis cells: Part I−Pure IrO_2-based anodes, Appl. Catal. B: Environ. 182 (2016) 153−160.

[138] F. Fouda-Onana, M. Chandesris, V. Médeau, S. Chelghoum, D. Thoby, N. Guillet, Investigation on the degradation of MEAs for PEM water electrolysers part I: Effects of testing conditions on MEA performances and membrane properties, Int. J. Hydrogen Energy 41 (2016) 16627−16636.

[139] M. Carmo, D.L. Fritz, J. Mergel, D. Stolten, A. Comprehensive, Review on PEM water electrolysis, Int. J. Hydrogen Energy 38 (2013) 4901−4934.

[140] E. Mayousse, F. Maillard, F. Fouda-Onana, O. Sicardy, N. Guillet, Synthesis and characterization of electrocatalysts for the oxygen evolution in PEM water electrolysis, Int. J. Hydrogen Energy 36 (2011) 10474−10481.

[141] C. Rozain, E. Mayousse, N. Guillet, P. Millet, Influence of iridium oxide loadings on the performance of PEM water electrolysis cells: Part II—Advanced oxygen electrodes, Appl. Catal. B: Environ. 182 (2016) 123−131.

[142] C. Rozain, P. Millet, Electrochemical characterization of polymer electrolyte membrane water electrolysis cells, Electrochim. Acta 131 (2014) 160−167.

[143] C. Rakousky, U. Reimer, K. Wippermann, S. Kuhri, M. Carmo, W. Lueke, et al., Polymer electrolyte membrane water electrolysis: restraining degradation in the presence of fluctuating power, J. Power Sources 342 (2017) 38−47.

[144] .Production d'hydrogène par électrolyse de l'eau (Mémento de l'Hydrogène), in AFHYPAC: Association française pour l'hydrogène et les piles à combustible, 2013.

[145] Ø. Ulleberg, Modeling of advanced alkaline electrolyzers: a system simulation approach, Int. J. Hydrogen Energy 28 (2003) 21−33.

[146] V.M. Rosa, M.B.F. Santos, E.P. Da Silva, New materials for water electrolysis diaphragms, Int. J. Hydrogen Energy 20 (1995) 697−700.

[147] M.M. Rashid, M.K. Al Mesfer, H. Naseem, M. Danish, Hydrogen production by water electrolysis: a review of alkaline water electrolysis, PEM water electrolysis and high temperature water electrolysis, Int. J. Eng. Adv. Tech. (IJEAT) 4 (2015) 2249−8958.

[148] T. Ohmori, K. Tachikawa, K. Tsuji, K. Anzai, Nickel oxide water electrolysis diaphragm fabricated by a novel method, Int. J. Hydrogen Energy 32 (2007) 5094−5097.

[149] K. Kinoshita, Electrochemcal Oxygen Technology, Wiley-Interscience, New York, 1992, pp. 78−99.

[150] M.E.G. Lyons, M.P. Brandon, The oxygen evolution reaction on passive oxide covered transition metal electrodes in aqueous alkaline solution. Part 1. Nickel, Int. J. Electrochem. Sci. 3 (2008) 1386−1424.

[151] A.J. Esswein, M.J. McMurdo, P.N. Ross, A.T. Bell, T.D. Tilley, Size-dependent activity of Co$_3$O$_4$ nanoparticle anodes for alkaline water electrolysis, J. Phys. Chem. C 113 (2009) 15068−15072.

[152] Y. Liang, Y. Li, H. Wang, J. Zhou, J. Wang, T. Regier, et al., Co$_3$O$_4$ nanocrystals on graphene as a synergistic catalyst for oxygen reduction reaction, Nat. Mater. 10 (2011) 780−786.

[153] M. Hamdani, M.I.S. Pereira, J. Douch, A. Ait Addi, Y. Berghoute, M.H. Mendonça, Electrochim. Acta 49 (2004) 1555−1596.

[154] R.N. Singh, D. Mishra, Anindita, A.S.K. Sinha, A. Singh, Novel electrocatalysts for generating oxygen from alkaline water electrolysis, Electrochem. Comm. 9 (2007) 1369−1373.

[155] H.-C. Chien, W.-Y. Cheng, Y.-H. Wang, T.-Y. Wei, S.-Y. Lu, Ultralow overpotentials for oxygen evolution reactions achieved by nickel cobaltite aerogels, J. Mater. Chem. 21 (2011) 18180−18182.

[156] D. Chanda, J. Hnát, M. Paidar, K. Bouzek, Evolution of physicochemical and electrocatalytic properties of NiCo2O4 (AB2O4) spinel oxide with the effect of Fe substitution at the A site leading to efficient anodic O2 evolution in an alkaline environment, Int. J. Hydrogen Energy 39 (2014) 5713−5722.

[157] H. Ichikawa, K. Matsuzawa, Y. Kohno, I. Nagashima, Y. Sunada, Y. Nishiki, et al., Durability and activity of modified nickel anode for alkaline water electrolysis, ECS Trans. 58 (2014) 9−15.

[158] M.H. dos Santos Andrade, M.L. Aciolia, J. Ginaldo da Silva Jr, J.C. Pereira Silva, E.O. Vilar, J. Tonholo, Preliminary investigation of some commercial alloys for hydrogen evolution in alkaline water electrolysis, Int. J. Hydrogen Energy 29 (2004) 235−241.

[159] J.M. Olivares-Ramírez, M.L. Campos-Cornelio, J. Uribe Godínez, E. Borja-Arco, R.H. Castellanos, Int. J. Hydrogen Energy 32 (2007) 3170−3173.

[160] D. Marcelo, A. Dell'Era, Daniel Marcelo, Alessandro Dell'Era, Int. J. Hydrogen Energy 33 (2008) 3041−3044.

[161] .Pătru A., Développement d'électrodes pour un électrolyseur alcalin H$_2$/O$_2$, PhD Thesis, Université de Montpellier II, 2013, pp. 39−52.

[162] M. Schalenbach, G. Tjarks, M. Carmo, W. Lueke, M. Mueller, D. Stolten, Acidic or alkaline? Towards a new perspective on the efficiency of water electrolysis, J. Electrochem. Soc. 163 (2016) F3197–F3208.

[163] A. Brisse, J. Schefold, M. Zahid, High temperature water electrolysis in solid oxide cells, Int. J. Hydrogen Energy 33 (2008) 5375–5382.

[164] M.A. Laguna-Bercero, Recent advances in high temperature electrolysis using solid oxide fuel cells: a review, J. Power Sources 203 (2012) 4–16.

[165] S. Li, R. Yan, G. Wu, K. Xie, J. Cheng, Composite oxygen electrode LSM.BCZYZ impregnated with Co3O4 nanoparticles for steam electrolysis in a proton-conducting solid oxide electrolyzer, Int. J. Hydrogen Energy 38 (2013) 14943–14951.

[166] X. Yang, J.T.S. Irvine, $(La_{0.75}Sr_{0.25})_{0.95}Mn_{0.5}Cr_{0.5}O_3$ as the cathode of solid oxide electrolysis cells for high temperature hydrogen production from steam, J. Mater. Chem. 18 (2008) 2349–2354.

[167] P. Moçoteguy, A. Brisse, A review and comprehensive analysis of degradation mechanisms of solid oxide electrolysis cells, Int. J. Hydrogen Energy 38 (2013) 15887–15902.

[168] K. Zeng, D. Zhang, Recent progress in alkaline water electrolysis for hydrogen production and applications, Prog. Energy Comb. Sci. 36 (2010) 307–326.

[169] A.S. Arico, S. Srinivasan, V. Antonucci, DMFCs: from fundamental aspects to technology development, Fuel Cell 1 (2001) 133–161.

[170] A. Serov, C. Kwak, Recent achievements in direct ethylene glycol fuel cells (DEGFC), Appl. Catal. B: Environ. 97 (2010) 1–12.

[171] R.B. De Lima, V. Paganin, T. Iwasita, W. Vielstich, On the electrocatalysis of ethylene glycol oxidation, Electrochim. Acta 49 (2003) 85–91.

[172] L. Demarconnay, S. Brimaud, C. Coutanceau, J.M. Léger, Ethylene glycol electrooxidation in alkaline medium at multi-metallic Pt based catalysts, J. Electroanal. Chem. 601 (2007) 169–180.

[173] Y.X. Chen, A. Lavacchi, H.A. Miller, M. Bevilacqua, J. Filippi, M. Innocenti, et al., Nanotechnology makes biomass electrolysis more energy efficient than water electrolysis, Nat. Commun. 5 (2014) 4036.

[174] T. Take, K. Tsurutani, M. Umeda, Hydrogen production by methanol–water solution electrolysis, J. Power Sources 164 (2007) 9–16.

[175] C.R. Cloutier, D.P. Wilkinson, Electrolytic production of hydrogen from aqueous acidic methanol solutions, Int. J. Hydrogen Energy 35 (2010) 3967–3984.

[176] G. Sasikumar, A. Muthumeenal, S.S. Pethaiah, N. Nachiappan, R. Balaji, Aqueous methanol eletrolysis using proton conducting membrane for hydrogen production, Int. J. Hydrogen Energy 33 (2008) 5905–5910.

[177] S. Kongjao, S. Damronglerd, M. Hunsom, Electrochemical reforming of an acidic aqueous glycerol solution on Pt electrodes, J. Appl. Electr. 41 (2011) 215–222.

[178] A. De Lucas-Consuegra, A.B. Calcerrada, A.R. De la Osa, J.L. Valverde, Electrochemical reforming of ethylene glycol. Influence of the operation parameters, simulation and its optimization, Fuel Process. Technol. 127 (2014) 13–19.

[179] F. Vigier, S. Rousseau, C. Coutanceau, J.M. Leger, C. Lamy, Electrocatalysis for the direct alcohol fuel cell, Top. Catal. 40 (2006) 111–121.

[180] S. Harish, S. Baranton, C. Coutanceau, J. Joseph, Microwave assisted polyol method for the preparation of Pt/C, Ru/C and PtRu/C nanoparticles and its application in electrooxidation of methanol, J. Power Sources 214 (2012) 33–39.

[181] A. Zalineeva, S. Baranton, C. Coutanceau, How do Bi-modified palladium nanoparticles work towards glycerol electrooxidation? An in situ FTIR study, Elecrochim. Acta 176 (2015) 705–717.

[182] Z. Hu, M. Wu, Z. Wei, S. Song, P.K. Shen, Pt-WC/C as a cathode electrocatalyst for hydrogen production by methanol electrolysis, J. Power Sources 166 (2007) 458–461.

[183] S.P. Sethu, S. Gangadharan, S.H. Chan, U. Stimming, Development of a novel cost effective methanol electrolyzer stack with Pt-catalyzed membrane, J. Power Sources 254 (2014) 161–167.

[184] W.J. Zhou, B. Zhou, W.Z. Li, Z.H. Zhou, S.Q. Song, G.Q. Sun, et al., Performance comparison of low-temperature direct alcohol fuel cells with different anode catalysts, J. Power Sources 126 (2004) 16–22.

[185] E. Antolini, Catalysts for direct ethanol fuel cells, J. Power Sources 170 (2007) 1–12.

[186] J.M. Léger, S. Rousseau, C. Coutanceau, F. Hahn, C. Lamy, How bimetallic electrocatalysts does work for reactions involved in fuel cells? Example of ethanol oxidation and comparison to methanol, Electrochim. Acta 50 (2005) 5118–5125.

[187] M. Simões, S. Baranton, C. Coutanceau, Electrooxidation of sodium borohydride at Pd, Au, and Pd_xAu_{1-x} carbon-supported nanocatalysts, J. Phys. Chem. C 113 (2009) 13369–13376.

[188] K. Okaya, H. Yano, H. Uchida, M. Watanabe, Control of particle size of Pt and Pt alloy electrocatalysts supported on carbon black by the nanocapsule method, ACS Appl. Mater. Interfaces 2 (2010) 888–895.

[189] H. Bönneman, W. Brijoux, R. Brinkmann, E. Dinjus, T. Joussen, B. Korall, Formation of colloidal transition metals in organic phases and their application in catalysis, Angew. Chem. Int. Engl. 30 (1991) 1312–1314.

[190] C. Xu, Y. Hu, J. Rong, S.P. Jiang, Y. Liu, Ni hollow spheres as catalysts for methanol and ethanol electrooxidation, Electrochem. Commun. 9 (2007) 2009–2012.

[191] V. Bambagioni, M. Bevilacqua, C. Bianchini, J. Filippi, A. Lavacchi, A. Marchionni, et al., Self-sustainable production of hydrogen, chemicals, and energy from renewable alcohols by electrocatalysis, ChemSusChem 3 (2010) 851–855.

[192] M. Simões, S. Baranton, C. Coutanceau, Electrochemical valorisation of glycerol, ChemSusChem 5 (2012) 2106–2124.

[193] M. Simões, S. Baranton, C. Coutanceau, Electro-oxidation of glycerol at Pd based nanocatalysts for an application in alkaline fuel cells for chemicals and energy cogeneration, Appl. Catal. B: Environ. 93 (2010) 354–362.

[194] H. Wang, B. Jiang, T.T. Zhao, K. Jiang, Y.Y. Yang, J. Zhang, et al., Electrocatalysis of ethylene glycol oxidation on bare and Bi-modified Pd concave nanocubes in alkaline solution: an interfacial infrared spectroscopic investigation, ACS Catal. 7 (2017) 2033–2041.

[195] Y. Kwon, Y. Birdja, I. Spanos, P. Rodriguez, M.T.M. Koper, Highly selective electrooxidation of glycerol to dihydroxyacetone on platinum in the presence of bismuth, ACS Catal. 2 (2012) 759–764.

[196] J. Gonzalez.Cobos, S. Baranton, C. Coutanceau, A systematic in situ infrared study of the electrooxidation of C3 alcohols on carbon-supported Pt and Pt–Bi catalysts, J. Phys. Chem. C 120 (2016) 7155–7164.

[197] J. Gonzalez-Cobos, S. Baranton, C. Coutanceau, Development of bismuth-modified PtPd nanocatalysts for the electrochemical reforming of polyols into hydrogen and value-added chemicals, ChemElectroChem 3 (2016) 1694–1704.

[198] W. Wang, Y. Kang, Y. Yang, Y. Liu, D. Chai, Z. Lei, PdSn alloy supported on phenanthroline-functionalized carbon as highly active electrocatalysts for glycerol oxidation, Int. J. Hydrogen Energy 41 (2016) 1272–1280.

[199] C. Coutanceau, A. Zalineeva, S. Baranton, M. Simoes, Modification of palladium surfaces by bismuth adatoms or clusters: effect on electrochemical activity and selectivity towards polyol electrooxidation, Int. J. Hydrogen Energy 39 (2014) 15877–15886.

[200] J.F. Gomes, A.C. Garcia, L.H.S. Gasparotto, N.E. de Souza, E.B. Ferreira, C. Pires, et al., Influence of silver on the glycerol electro-oxidation over AuAg/C catalysts in alkaline medium: a cyclic voltammetry and in situ FTIR spectroscopy study, Electrochim. Acta 144 (2014) 361–368.

[201] C.A. Ottoni, S.G. da Silva, R.F.B. De Souza, A.O. Neto, Glycerol oxidation reaction using PdAu/C electrocatalysts, Ionics 22 (2016) 1167–1175.

[202] Q. Lin, Y. Wei, W. Liu, Y. Yu, J. Hu, Electrocatalytic oxidation of ethylene glycol and glycerol on nickel ion implanted-modified indium tin oxide electrode, Int. J. Hydrogen Energy 42 (2017) 1403–1411.

[203] V.L. Oliveira, C. Morais, K. Servat, T.W. Napporn, G. Tremiliosi-Filho, K.B. Kokoh, Glycerol oxidation on nickel based nanocatalysts in alkaline medium—identification of the reaction products, J. Electroanal. Chem. 703 (2013) 56–62.

[204] M.V. Pagliaro, M. Bellini, M. Bevilacqua, J. Filippi, M.G. Folliero, A. Marchionni, et al., Carbon supported Rh nanoparticles for the production of hydrogen and chemicals by the electroreforming of biomass-derived alcohols, RSC Adv. 7 (2017) 13971–13978.

[205] A. Zalineeva, A. Serov, M. Padilla, U. Martinez, K. Artyushkova, S. Baranton, et al., Self-supported PdxBi catalysts for the electrooxidation of glycerol in alkaline media, J. Am. Chem. Soc. 136 (2014) 3937–3945.

[206] A. Zalineeva, A. Serov, M. Padilla, U. Martinez, K. Artyushkova, S. Baranton, et al., Glycerol electrooxidation on self-supported Pd1Snxnanoparticules, Appl. Catal. B: Environ. 176 (2015) 429–435.

[207] A. Caravaca, F.M. Sapountzi, A. de Lucas-Consuegra, C. Molina-Mora, F. Dorado, J.L. Valverde, Electrochemical reforming of ethanolewater solutions for pure H_2 production in a PEM electrolysis cell, Int. J. Hydrogen Energy 37 (2012) 9504–9513.

[208] A.T. Governo, L. Proença, P. Parpot, M.I.S. Lopes, I.T.E. Fonseca, Electro-oxidation of d-xylose on platinum and gold electrodes in alkaline medium, Electrochim. Acta 49 (2004) 1535–1545.

[209] J.A. Turner, Sustainable hydrogen production, Science 305 (972) (2004) 972–974.

[210] R. Kothari, D. Buddhi, R.L. Sawhney, Comparison of environmental and economic aspects of various hydrogen production methods, Renew. Sustainable Energy Rev. 12 (2008) 553–563.

[211] A. Ilie, M. Simoes, S. Baranton, C. Coutanceau, S. Martemianov, Influence of operational parameters and of catalytic materials on electrical performance of direct glycerol solid alkaline membrane fuel cells, J. Power Sources 196 (2011) 4965–4971.

[212] P.T. Anastas, J.C. Warner. In Green Chemistry: Theory and Practice, Oxford University Press, Oxford, UK, (2000). ISBN: 0198506988, 9780198506980.

[213] Brundtland Report, "Our common future", World Commission on Environment and Development, United Nation Organization (UNO), New York, 1987.

INDEX

Note: Page numbers followed by "*f*" and "*t*" refer to figures and tables, respectively.

Printed in the United States
By Bookmasters